ANALYSIS AND DESIGN OF ADVANCED MULTISERVICE NETWORKS SUPPORTING MOBILITY, MULTIMEDIA, AND INTERNETWORKING

Analysis and Design of Advanced Multiservice Networks Supporting Mobility, Multimedia, and Internetworking

COST Action 279 Final Report

Edited by

JOSE BRAZIO

PHUOC TRAN-GIA

NAIL AKAR

ANDRZEJ BEBEN

WOJCIECH BURAKOWSKI

MARKUS FIEDLER

EZHAN KARASAN

MICHAEL MENTH

OLIVIER PHILIPPE

KURT TUTSCHKU

SABINE WITTEVRONGEL

COST Action 279

A C.I.P. Catalogue record for this book is available from the Library of Congress.

ISBN 13 978-1-4419-3927-2
ISBN 10 0-387-28173-8 (e-book)
ISBN 13 978-0-387-28173-5 (e-book)

Published by Springer,
P.O. Box 17, 3300 AA Dordrecht, The Netherlands.

www.springeronline.com

Printed on acid-free paper

Printed in the Netherlands.

Preface

This book constitutes the Final Report of COST Action 279, *Analysis and Design of Advanced Multiservice Networks supporting Multimedia, Mobility, and Interworking*, a guided tour of the state-of-the-art work on diverse aspects of modern telecommunications networks design developed within this Action during the four years of its operation, started on July 1, 2001, and ended on June 30, 2005.

As stated in its founding charter, its Memorandum of Understanding, the work area of COST 279 is the analysis, design, and control aspects of present-day networks—quite a wide scope. Behind the unifying façade put on by the Internet Protocol (IP) network layer, todays networks hide a mess of heterogeneity: heterogeneity at the level of applications, both concerning the traffic they produce and the network Quality of Service (QoS) they require, and heterogeneity at the level of network component subsystems, in particular an increasingly important mobile/wireless access segment. A common ground for the treatment of this disparate set of topics was given by the strong methodological component contained in the approach followed in COST 279, with importance placed on the development and application, whenever possible, of analytical techniques and models for the mathematical understanding of the systems under study. The results expected from the Action ranged thus from mathematical models and algorithms as entities of own interest to the understanding of system behavior via their application. Explicit value was also given to contributions to progress in basic issues, such as queueing theory and estimation and identification in stochastic models, in recognition of their status as key elements for the understanding of the behavior of networks and for their design.

During its period of operation, COST 279 was a crucible for a very fruitful interaction among a sizeable group of European researchers exhibiting a remarkable diversity across a number of dimensions: with origins from both telecom operators and universities, with backgrounds in mathematics, computer science, and engineering, and with experiences ranging from basic research to applied engineering. The competences available in this group thus matched quite well the diversity accommodated within the scope of the Action.

The current book summarizes the work contained in the internal technical documents, officially designated *Temporary Documents*, presented in the meetings held during the lifetime of COST 279. The work in some of those TDs has since been published in the peer-reviewed literature, but the work in some others, especially the more recent ones, has not. In the text of the following chapters, references are made to the open literature publications, whenever possible. When only the internal TDs are available, a reader looking for further details is asked to have the kindness of contacting directly the authors for the materials of his interest. The list of author contacts and TDs produced, together with their abstracts, is available at the Action web site, at `http://www.lx.it.pt/cost279/`.

Given its origin and purpose, the present publication is not a textbook. It is rather an annotated bibliography on a body of state-of-the-art work done on a related set of topics on network design, and it has an intrinsic value as such. However, a reader wanting to get an overall picture of the state-of-the-art in a specific sub-area within the scope of COST 279 can most likely do so by systematically exploring the numerous references given in the text.

The book starts with a short *Introduction* to the COST Research Framework and to COST Action 279 itself. The rest of the book, containing the technical material proper, has a hybrid organization. The first technical chapter is on *IP-Based Networks*, and deals essentially with issues relevant to end-to-end QoS. The following two chapters are of a horizontal nature, the first on *Queueing Models*, and the second on *Traffic Measurement, Characterization, and Modeling*. The last three chapters are technology-specific, and cover *Wireless Networks*, *Optical Networks*, and *Peer-to-Peer Services*. The book contains, as Appendices, an extensive *Bibliography*, the *List of Temporary Documents*, and the details on the *COST 279 Management Committee* and the *COST 279 Participating Institutions*.

Because of the hybrid structure of the book, the study of specific topics may fit naturally into more than one chapter, particularly so when both a basic, methodological part, and a system study part are involved. In such cases, the option made was to include references to the work in both chapters, with appropriate cross-references. In this way, chapters are, as much as possible, self-contained.

This book would not have been possible without the contribution of an enthusiastic and hard-working team of people. First and foremost, there are all the members of COST 279, the people who did the technical work reported, and who are indeed the reason for its existence. A large debt is next owed to Michael Menth, from the University of Würzburg, Germany, who single-handedly coordinated the organization of the *COST 279 Mid-Term Report*, upon which this book is based, together with the fine team of technical chapter editors and contributing members listed at the beginning of each chapter. The overall operation of COST 279 had the wise guidance of its Management Committee and the support of the COST Program Technical Committee for Telecommunications and Information Science and Technology and the COST TIST Secretariat.

In the course of the year 2003 we had the sad news of the passing away of Prof. Olga Casals, MC member from Spain. COST 279 remains indebted to both her technical contributions and personal enthusiasm.

Even though COST 279 ended at its normal four-year term, the ensemble of its members possesses high enough cohesion and momentum to allow itself to materialize in the future, given appropriate conditions, into a similar initiative. We remain in anticipation of its outcome.

June 2005

José Brázio
Chairperson

Phuoc Tran-Gia
Vice-Chairperson

Coordination

COST Action 279 Management Committee

José Brázio, Chairman

Phuoc Tran-Gia, Vice-Chairman

Chapter Editors and Contributors[1]

IP-Based Networks

Editors: M. Menth, P. Olivier

Contributors: M. Çaglar, I. Gojmerac, M. Klimo, M. Menth, M. Meo, S. Molnár, I. Norros, P. Olivier, S. Oueslati, Ö. Özkasap, H. Tarasiuk

Queueing Models

Editor: S. Wittevrongel

Contributors: S. Aalto, N. Akar, C. Belo, A. da Silva Soares, S. De Vuyst, M. Fiedler, D. Fiems, P. Gao, V. Inghelbrecht, R. Janowski, U. Krieger, K. Laevens, T. Maertens, M. Mandjes, I. Norros, O. Østerbø, D. Sass, K. Spaey, H. Tran, H. van den Berg, J. Virtamo, J. Walraevens, S. Wittevrongel

Traffic Measurement and Characterization

Editor: M. Fiedler

Contributors: P. Arlos, M. Çaglar, M. Fiedler, G. Hu, J. Kilpi, U. Krieger, P. Mannersalo, S. Molnár, L. Muscariello, P. Olivier, Ö. Özkasap, A. Popescu, M.-A. Remiche, K. Salamatian, P. Salvador, K. Tutschku, H. van den Berg, S. Wittevrongel

Wireless Networks

Editors: W. Burakowski, A. Beben

Contributors: S. Aalto, H. van den Berg, R. Boucherie, Ll. Cerdà, B. Zovko-Cihlar, M. Fiedler, R. de Haan, T. Hoßfeld, E. Hyytiä, S. Imre, L. Isaksson, J. Kilpi, R. Litjens, M. Meo, M. Mohorčič, R. Pries, F. Roijers, D. Staehle, A. Švigelj, M. Szalay, A. de Vendictis

Optical Networks

Editors: E. Karasan, N. Akar

Contributors: D. Fiems, C. Gauger, G. Hu, E. Hyytiä, V. Inghelbrecht, K. Laevens

Peer-to-Peer Services

Editor: K. Tutschku

Contributors: A. Binzenhöfer, H. Reittu. M. L. García Osma, Ö. Özkasap, K. Tutschku

[1] Please see chapters for detailed affiliations.

Contents

Introduction

The COST Program

COST 279 is one of the Actions of the European COST Program, an intergovernmental framework for European Co-operation in the field of Scientific and Technical Research allowing the co-ordination of nationally funded research on a European level. COST Actions are launched on a "bottom-up" approach, with the initiative for the creation of Actions coming from the scientists and technical experts themselves and from those with a direct interest in furthering international collaboration. According to the general spirit of flexibility of COST, country participation follows an *à la carte* principle.

The COST framework provides means for the setting up of regular meetings among researchers of the participating countries, for the purpose of technical exchanges, discussion of research directions, and organization of common initiatives. The resulting co-operation and interaction among researchers is intended to help Europe hold a strong position in the field of scientific and technical research. Its experience has shown very beneficial to the research community: besides helping keep its cohesion, it has allowed advanced students and young researchers in the beginning of their careers to get in touch with the latest research developments and integrated into the community, and it has potentiated the start up of research groups in environments where previous research tradition in an area did not exist. In addition, COST provides mechanisms for the transfer of its research results to the surrounding society, and examples abound of related "success stories. Detailed information on the COST Program can be obtained via the Web at the URL `http://www.cordis.lu/cost/`.

COST Action 279

COST Action 279 belongs to a distinguished lineage of COST Actions (COST 201, 214, 224, 242, and 257) that, since 1979, have fostered cooperation

J. Brazio et al. (eds.), Analysis and Design of Advanced Multiservice Networks Supporting Mobility, Multimedia, and Internetworking, 1–4.

among European researchers in the field of teletraffic and multiservice network performance and design. This sequence of Actions has been carried on by what can be called a core group of the European teletraffic community, its members originating from industry, operators, and universities, and has been an ever-present actor in all the steps of the evolution of modern (digital) telecommunications networks—from circuit-switched networks to the introduction of ISDN and ATM in the 1980s and on to todays drive towards all-IP networks, from fixed to mobile and wireless service, from copper-based to fiber-optical transmission, and from electronic to optical switching. It thus has all the richness of the memory and experience of the past, together with the promise and potential for the future of its active members and permanently joining new researchers.

The Memorandum of Understanding that established COST 279 states: "The main objective of the Action is to develop techniques for the analysis, design and control of advanced multiservice networks supporting mobility, multimedia and interworking, by means of the development and application of new and better analytical techniques for the mathematical understanding and optimisation of the behaviour of communications equipment, protocols, and network topologies and architectures, and of economic aspects such as pricing principles and network cost estimation. The results will have the form of mathematical models and results, algorithms, computer tools, and analyses of empirical traffic and network data. The Action is also expected to contribute to general, theoretical progress in basic issues like queueing theory, estimation and identification in stochastic models, and simulation, in particular as applied to rare events."

For the achievement of its objectives the COST 279 community recognized the importance of methodological progress regarding the analysis, design and planning of multiservice networks supporting advanced requirements. Therefore, a main distinguishing feature of the scientific program of the Action resided on the strong emphasis on modeling and performance issues, as pertaining to both the basic support disciplines and to their application to the study of the behavior of and interaction among networks, end-systems, services, and user applications.

COST 279 started on July 1, 2001, at a time where the research on ATM had been phased-out, and had been phased-in the research resulting from the widespread interest, arisen a few years before, of the use of the Internet Protocol for the operation and interconnection of heterogeneous networks. The Action steadily contributed since that time to the understanding of these systems and provided inputs for their design, by means of work on topics such as techniques for traffic measurement, characterization, and modeling, network dimensioning, control and engineering, and support of Quality-of-Service, all of these in the context of wired, wireless, and optical networks and their interconnection. COST 279 was also swift in adapting to the appearance of new topics within its scope, even if not initially foreseen in the original workplan, as shown by the existence of the *Peer-to-Peer Services* chapter in this book. In parallel, the Action continued to produce developments in basic issues like queuing models, statistical estimation, and random graph theory. The collection of these results is embodied in the 173 internal documents listed in the Appendix *List of Temporary Documents* (TDs) at the end of this book. From the work contained in these TDs, about 155 publications have resulted at the time of the publication of this book in the peer-reviewed literature, the references of which are included in the Appendix *Bibliography*

The core of the operation of COST 279 took place at its Management Committee (MC) Meetings, occurring regularly at four-month intervals. In addition to the administrative and policy management aspects of the Action, the meetings included technical sessions where the internal technical documents were presented and discussed in an informal environment, with room for degrees of work maturation going from exploratory ideas, to work in progress, and all the way up to finished and published work. The composition of the MC is given in the corresponding Appendix. The meetings that took place were attended by 170 different participants, coming from 50 research institutions from 23 countries, also listed in appendices at the end of the book.

In addition to its "internal" operation, COST 279 implemented activities to transfer its know-how to the outside community. One such activity was the organization of two Seminars, at the mid- and end-points of its operation, where its results were publicly presented. The Mid-Term Seminar took place in Rome, Italy, on January 23–25, 2004, with local organization provided by Andrea Baiocchi, from the University of Rome "La Sapienza." The Final Seminar took place in Lisbon, Portugal, on June 27–29, 2005, having as local organizer José Brázio, from the Telecommunications Institute at Instituto Superior Técnico. Another activity was the implementation of Summer Schools, specifically targeted for qualifying advanced students and young researchers of European research institutions in the field, with the stated objective of contributing to further strengthening the joint European research community. The first COST 279 Summer School, on *Stochastic Modeling and Analysis in Telecommunication* with emphasis on *Wireless Systems*, took place in August 2002, and was attended by 32 doctoral students from 20 research institutions belonging to 15 COST countries. The second edition of the Summer School took place in September 2003 having as topic *Routing and Multi-Layer Traffic Engineering in Next Generation IP Networks*, and was attended by 34 students coming from 27 research institutions of 18 COST countries. The Summer Schools took place at the Conference Center of Deutsche Telekom in Darmstadt, Germany, and were the brainchild of Udo Krieger, now at the Otto Friedrich University Bamberg, Germany, who ran all its logistical and technical organization aspects. In both editions, lectures were provided by top European experts in the field, pertaining to both universities and telecommunications operators. The initiative of Summer Schools, not frequent among COST Actions at the time, started as a learning experiment that was quickly recognized to be a success.

Chapter 1
IP-Based Networks

Michael Menth
University of Würzburg, Germany

Philippe Olivier
France Telecom, France

Contributors:
Mine Çaglar (Koç University, Turkey), Ivan Gojmerac (FTW Vienna, Austria), Martin Klimo (University of Zilina, Slovakia), Michael Menth (University of Würzburg, Germany), Michela Meo (Politecnico di Torino, Italy), Sándor Molnár (Budapest University of Technology and Economics, Hungary), Ilkka Norros (VTT, Finland), Philippe Olivier (France Telecom, France), Sara Oueslati (France Telecom, France), Öznur Özkasap (Koç University, Turkey), Halina Tarasiuk (Warsaw University of Technology, Poland)

1.1 Introduction

The Internet provides interconnection of networks based on the Internet Protocol (IP). In contrast to the structure of the telephone network, circuit-switched and connection-oriented, the Internet is packet-switched and connectionless. The basic service provided by IP is unreliable in what concerns packet delivery. At the level of the transport service available to applications, the Internet provides both an unreliable transport service, supported on the simple User Datagram Protocol (UDP), and a reliable transport service supported on the Transmission Control Protocol (TCP). TCP peers are located only at the end systems. The design of the Internet thus leaves its core stateless and the network unaware of higher protocol layers, e.g., the service layer. Therefore, it is simpler to deploy new services in the Internet than in the telephone network. The telephone world, however, has a revenue creating property not provided by IP networks: Quality of Service (QoS). This feature must be offered by a next generation of the Internet if circuit-switched networks are to be integrated in or replaced by packet-switched networks.

The QoS feature can be implemented, e.g., by Admission Control (AC), Capacity Overprovisioning (CO), or class-based services such as Differentiated Services (DiffServ), not necessarily excluding each other. Admission Control requires AC entities in the network, raising interoperability issues and

J. Brazio et al. (eds.), Analysis and Design of Advanced Multiservice Networks Supporting Mobility, Multimedia, and Internetworking, 5–54.

causing upgrade costs for systems. It also requires state storage in the networks, thus introducing the possibility for failures and potentially compromising the network scalability. These problems are to some extent avoided by CO but this approach has also some disadvantages: it leads to permanent increased bandwidth costs and the needed overdimensioning factor is not known; moreover, we may argue that, while overprovisioning is designed to meet reasonably predictable traffic variations, some kind of AC is still needed to react to some network failures or unpredicted traffic rushes, e.g., due to exceptional events. Hence, the technical future for the Next Generation Network (NGN) is not yet decided and some of the research carried out in COST 279 contributes to the decision-making process in the sense that all directions are investigated.

Some of the work on the topic under COST 279 was done in three major research and development projects for NGNs. Results from these projects are partly described in the first three sections of the present chapter.

- AQUILA (Adaptive resource control for QoS Using an IP-based Layered Architecture) was a European research project partially funded by the Information Society Technologies (IST) priority of the European Commission Fifth Framework Programme. AQUILA had a 39 months duration, from January 2000 to March 2003. Its principal aim was to define a QoS architecture for IP networks, mainly based on the DiffServ concepts. More details are given on the Website `http://www-st.inf.tu-dresden.de/aquila/`.
- KING (Key components for the Internet of the Next Generation) was a project partially funded by the German Ministry for Education and Research. Siemens Information and Communication Networks ran the project from October 2001 to March 2005, together with a number of German universities and research institutes. The overall objective of KING was to develop efficient solutions for carrier-grade IP networks that satisfy high QoS and resilience requirements [1], while at the same time providing low operational overheads. The Website can be found at `http://www.siemens.com/king/`.
- EuroNGI (Design and Engineering of the Next Generation Internet) is a Network of Excellence funded by the IST priority under the European Commission Sixth Framework Program. Gathering almost sixty research institutions from most of the european countries, it started in December 2003 and is planned to have a 3 years duration. Within this project, the Joint Research Activity 2 particularly provides work packages whose general objectives are on traffic engineering and management, congestion control, and end-to-end QoS in multiservice networks.

The entry page of the EuroNGI Website can be found at
`http://eurongi.enst.fr/`.

The structure of this chapter is as follows. First, Section 1.2 proposes different service models to provide QoS in IP-based networks, based either on strict QoS, relative QoS or implicit service differentiation. Then, Section 1.3 deals with practical applications of well-known results for AC and shows their feasibility. So far, AC has been investigated for unresponsive traffic, e.g., real-time traffic, whereas AC approaches for elastic TCP traffic can adapt to bottlenecks. Section 1.4 extends the concept of AC from Link AC (LAC) for a single link to Network AC (NAC), which addresses the storage of AC information, i.e., the reservation states. Since failure recovery is harder for stateful networks, resilience issues are studied in this context.

Section 1.5 is dedicated to the dimensioning of network links. Today's IP networks have to be provisioned reasonably to get satisfying QoS values for elastic traffic, e.g., response time for TCP-based applications such as Web browsing, file transfer, or file sharing. Then, TCP enhancements, as well as some issues on packet level performance, are considered in Section 1.6; all of them contribute to the necessary adaptation of the TCP-UDP/IP protocol family to the NGN and its new requirements in terms of QoS-oriented control mechanisms.

The following sections of this chapter are dedicated to network improvements. Scheduling mechanisms are described in Section 1.7. Such mechanisms, aiming at decreasing the queuing time for delay sensitive data, are typically required in the framework of DiffServ-like service architectures. Routing optimization and load balancing, addressed in Section 1.8, avoid overload and thereby improve QoS or, conversely, allow network operation at a given QoS level with less resources. Concurrently, the multicast technology, which forms the subject of Section 1.9, contributes to network efficiency because of the reduction in traffic volume brought about by the sharing of transmission resources.

Finally, results obtained for most of the above-mentioned issues may depend on the network topology. Therefore, Section 1.10 reports on some meaninful advances recently achieved in network graph modeling, which help understanding the topology of the global Internet.

1.2 Service Models for QoS IP Networks

Modern QoS IP networks are targeted for effective handling of traffic produced by a variety of application types currently available to the Internet users.

These applications differ in transmission requirements, demanding from the network the support of a number of network services differing in QoS objectives. The design process of a network service assumes that a specific traffic profile should be handled by the network with a predefined QoS, usually expressed by such parameters as packet transfer delay and packet loss ratio. To meet these requirements, some QoS mechanisms should be implemented at different levels in the network. Mechanisms like classifiers, conditioners and schedulers are used at the packet level, while AC procedures are performed at the flow level.

We first classify QoS architectures with regard to *strict* or *relative* QoS. For providing strict QoS, the AC function constitutes a key element, allowing the regulation of the volume of traffic submitted to a network with limited resources. On the contrary, the relative (proportional) QoS does not use AC but rather assumes that high priority flows are handled by the network in a preferred way. This section contains a brief overview of network service types that are recognized as adequate for networks with strict and relative QoS, including an extension of these concepts to provide End-to-End QoS. Then an alternative QoS architecture is proposed, named as Cross-Protect, which is based on implicit service differentiation and Measurement-Based Admission Control (MBAC) at the flow level.

1.2.1 Network Services with Strict QoS

A framework for providing QoS differentiation and strict QoS guarantees in IP networks is developed in the AQUILA project [2]. This approach assumes network support for four premium network services and is based on the DiffServ concept. Each network service is designed to support a class of applications with similar QoS requirements and traffic characteristics. For each premium network service, an adequate Traffic Descriptor-based AC (TDAC) method is proposed based on AC algorithms developed for Asynchronous Transfer Mode (ATM) networks.

- AC for the Premium Constant Bit Rate (PCBR) network service is based on the Rate Envelope Multiplexing (REM) scheme using a well-known peak rate allocation method. The admissible load for the capacity allocated to PCBR is calculated based on the analysis of an M/D/1/K queueing system depending on the assumed target packet loss ratio and on the available buffer size. PCBR is dedicated to Constant Bit Rate (CBR) traffic, e.g., voice trunks, and is served with the highest priority. Negligible packet delay variation can be assured when at most 10% PCBR load is allowed on the link.

- The REM scheme is also applied to the Premium Variable Bit Rate (PVBR) network service, which is designed for the effective transfer of streaming Variable Bit Rate (VBR) traffic, e.g., video applications. Here, the effective bandwidth is calculated by Lindberger's method [3]. The buffer dimensioning rules are adequate to absorb the so-called packet-scale congestion. For this purpose, the analysis of the N.D/D/1 queueing system is applied.

- The Premium Multi-Media (PMM) network service is used for greedy TCP, or TCP-like, flows. Its AC method is a function of the sustained bit rate, a value which corresponds to the minimum flow rate that is required.

- The Premium Mission Critical (PMC) network service is conceived to assure few losses for elastic non-greedy flows. Its AC method adapts the Rate Sharing Multiplexing (RSM) scheme and the effective bandwidth calculation.

- Finally, the Standard network service is designed for carrying best effort traffic.

An extension of MBAC methods for streaming flows and two novel AC concepts for the PMM service are provided in [4, 5]. The final results of the AQUILA project confirm the expected efficiency of the proposed AC methods for the considered premium network services. They show the feasibility of differentiated QoS network services by adding new functionalities into existing IP networks. Some selected results are presented, e.g., in [6].

1.2.2 Network Services with Relative QoS

A solution for providing relative QoS using DiffServ mechanisms is proposed in [7]. The authors choose the Dynamic Real-Time/Non Real-Time (RT/NRT) mechanism which has received less attention than Assured Forwarding (AF), Expedited Forwarding (EF), or the related Simple Integrated Media Access (SIMA) proposal [8]. SIMA relies both on financial and end-to-end congestion control incentives to achieve differentiated bandwidth allocation. The entity that influences both charging and bandwidth division is the nominal bit rate. AC is not considered in this scheme. First, flow and packet level models for SIMA are presented in [7] and compared for elastic traffic. Then, fair differentiation among elastic flows using SIMA is shown as well as fair differentiation between TCP and non-TCP flows. Finally, the applicability of SIMA as a future Internet Service model is demonstrated. The obtained results illustrate that

SIMA does not produce any fixed weights, determined by the nominal bit rate or price paid, between bandwidth allocations for the different classes. However, the provided models show that SIMA does achieve class differentiation, where the weights may vary from equal allocation to allocation in proportion of the nominal bit rate purchase.

1.2.3 Network Services for End-to-End QoS

Nowadays, we observe a marked evolution and diversification of access networks, such as Ethernet, xDSL (x Digital Subscriber Line), WLAN (Wireless Local Area Network), and UMTS (Universal Mobile Telecommunications System). The interconnection of such networks via the IP core network constitutes a heterogeneous environment, particularly from the QoS support point of view. The problem of ensuring end-to-end QoS in heterogeneous networks is investigated in [9]. Providing QoS in such an environment requires the definition of network services and dimensioning of network resources for intra- and inter-domain areas. In [9] is introduced the network service concept and investigated the quality of end-to-end packet transfer over the heterogeneous network supporting such a concept. It is stated that one needs to specify end-to-end network services which will be visible by the applications (end users). The packet traffic from an application is submitted to one of the end-to-end network services, and the experienced QoS at the packet level straightforwardly depends on the effectiveness and reliability of the service. A list covering all currently discussed end-to-end services is presented in [10]. In [9], the authors follow this concept and distinguish between a maximum of eleven end-to-end network services differing in both pre-defined QoS objectives and traffic profiles. On the contrary, based on discussions between service providers, the authors in [10] propose only four inter-provider network services: (i) *Ctrl*, for transferring system information, (ii) *Real-time*, for transferring so-called streaming traffic, sensitive to packet loss and delay, (iii) *Non Real-Time*, for transferring packet streams non-sensitive to packet loss and delay, but requiring a guaranteed transmission rate, and (iv) *Best Effort*, for the rest of traffic.

Taking into account the above, the authors also investigate the following mapping functions: (i) a function between applications and end-to-end network services, (ii) a function between end-to-end network services and network services available in particular network technologies, such as WLAN, Ethernet, and IP, (iii) a function between end-to-end network services and inter-provider network services.

The paper also aims at providing an adequate formula for obtaining the admissible traffic load when one maps two end-to-end network services dedicated to Voice over IP (VoIP) (Telephone Service) and video conference (MM Conferencing Service), respectively, into one inter-provider network service, which is *Real-time*. In this case, the QoS objectives of Real-time service are very low values of packet delay, jitter and packet loss. The analysis in [9] determines the admissible load when the target packet loss and buffer size dedicated to the Real time service are known. The proposed solution includes a study of the different packet sizes in voice streams, about 100 bytes, and in video streams, about 1500 bytes. As a first approach, the admissible load is obtained by applying the M/D/1/K queue analysis. However, as expected, simulation results show that, when two Poisson packet streams of different, but constant, packet sizes share a common buffer, dimensioned in packets, the packet loss probability essentially depends on the ratio of these packet sizes. Increasing the packet size of one stream, while maintaining the same traffic load and packet size of the other stream, increases the packet loss ratio. The authors provide a relatively simple mathematical model that accounts for this phenomena, by considering a discrete time queue as in [11], where time is slotted into units corresponding to the smaller packet size. By applying the method, the *Real-time* service can meet the targeted packet loss ratio in a variety of traffic parameter settings (packet size ratio, proportion of the contributing loads) by appropriately adjusting the maximum admissible load, even when the packet sizes differ significantly. The effectiveness of the proposed method is illustrated by simulation results.

1.2.4 Implicit Service Differentiation

In [12], an MBAC mechanism is combined with a Priority Fair Queueing (PFQ) scheduler that implicitly differentiates between elastic and streaming traffic in the context of an integrated QoS architecture. The name of the proposed architecture, *Cross-protect*, indeed refers to this combination of mechanisms. Obviously, elastic and streaming traffics require differentiated handling due to their different QoS needs. The PFQ discipline is based on the Start-time Fair Queueing (SFQ) algorithm [13], which imposes Max-Min fairness at the flow level. Implicit differentiation is realized by giving priority to packets of flows whose rate is less than the current fair rate. The residual bandwidth is fairly shared by the other flows, referred to as bottlenecked flows. In contrast to other service differentiation models, such as DiffServ, which rely on packet marking or packet classification according to explicit predefined rules, *Cross-protect* uses the current rate characteristics of flows to differentiate between streaming and elastic traffic.

Implicit flow level AC, particularly for elastic traffic, will be described in more details in Section 1.3.2. In *Cross-protect*, the role of AC is to enforce a minimum fair rate which, on one hand, is greater than the peak rate of most streaming applications and, on the other hand, represents the minimum throughput guaranteed to elastic flows. In the meanwhile, streaming flows are assured to be given priority treatment. Hence PFQ, along with AC, ensures negligible loss and delay for streaming flows under all load conditions, and imposes a minimum throughput for unconstrained elastic flows. In this proposal, the AC module uses two measurements provided by the PFQ scheduler, the *fair rate* and the *priority load*. The fair rate estimates the rate obtainable by a fictitious permanent bottlenecked flow, and the priority load represents the load induced by the traffic handled with priority. New flows are rejected if the estimated fair rate drops below a predefined threshold, or if the priority load exceeds another predefined threshold. Both thresholds need to be compatible with QoS requirements and efficient resource utilization.

Such an approach as *Cross-protect* relies, particularly, on the scalability of per-flow fair queueing. It is shown in [14] that, contrary to what is usually believed, the deployment in the Internet of per-flow fair queuing based on per-flow state maintenance is feasible, despite the extremely high number of flows which may be present on high speed links. The basic idea consists of distinguishing between flows in progress and active flows, the latter being defined as those flows actually having packets in the queue. The authors show that, while the number of flows in progress increases with the link rate, the number of active flows typically remains limited to hundreds of flows and does not significantly vary with the link rate. Indeed, the number of active flows, for which state maintenance is needed in the scheduler, is limited to the flows which are bottlenecked at the queue and to a small proportion of non bottlenecked flows. The study is performed by running simulations based on real Internet traces, under three scenarios representative of realistic and common situations, and by developing analytical models explaining the queue behavior.

1.3 Admission Control in IP Networks

In the last decade, many AC methods have been developed for traffic control in ATM networks. Both ATM and IP networks are packet-switched technologies with many similarities. Therefore, it seems natural to adopt the AC methods for ATM to IP technology, but not without considering their differences. Some important issues are:

- IP networks require a higher level of preventive admission control than ATM networks, because the variable size of IP packets induces higher jitter and delay than the fixed-size cells of ATM networks. For instance, small voice packets of 60 bytes may experience relatively large delay when they wait for the end of transmission of a 1500 bytes data packet, thus leading to serious problems for the efficient multiplexing of traffic streams with different profiles.

- Effective AC mechanisms were developed for streaming traffic in ATM. These results are not applicable to elastic TCP-controlled traffic, thus new AC methods are required for IP networks.

AC methods can be classified into two main types: TDAC methods and MBAC methods. TDAC methods can be again split into deterministic and statistical multiplexing schemes. Furthermore, both statistical and MBAC methods may refer to REM or to RSM. The above classification, earlier introduced for ATM, can be extended to IP.

1.3.1 Admission Control for Streaming Traffic

In the following we present a number of studies dedicated to AC for streaming traffic, typically created by audio/video applications: an estimation of the admissible load in a two-priority system; a two-stage AC method for voice traffic; an MBAC approach for non-real-time VBR traffic; and an application of AC results to investigate protocol design alternatives for low bit rate real-time traffic. These studies are based on well-known AC methods for ATM networks, which are here extended to take into account the specific properties of IP networks.

AC for a Two-Priority System

The AQUILA architecture, based on the DiffServ concept, assumed Priority Queueing (PQ) and Weighted Fair Queueing (WFQ) mechanisms at the edge routers to differentiate service facilities among defined network services (traffic classes) [2]. For that purpose, an amount of resources, bandwidth and buffer size, is dedicated to each traffic class. However, up to now, the impact of higher priority traffic on the admissible load for low priority classes has not been considered.

Therefore, in [15], a method is proposed to determine the maximum admissible load for the low priority class in a two-priority system with a non-preemptive priority scheduler. The method derives the probability distribution of the low priority queue length which, given the buffer size and the required packet loss ratio (i.e., the targeted QoS parameter), allows the determination of the admissible load for the low priority service. Besides, the impact of total traffic load and packet sizes of both priority classes is studied. More details on the approach, mainly based on a single queue transformation and on a diffusion approximation, and on the obtained results can be found in Section 2.7 of the Queueing Models Chapter. Because of the assumption of Poissonian input, the method is readily applicable to the superposition of an infinite number of CBR streams. However, it is also applicable to other input processes with finite mean and variance.

A Two-Stage AC Method

As mentioned before, IP networks serve packets with different sizes. This feature of IP traffic introduces additional problems for delay and jitter guarantees for streaming (real-time) traffic as compared to ATM networks. A possible solution has been proposed in [16]. A two-stage Connection Admission Control (CAC) model based on the REM and Simplified Reference Model (SiRM) concepts is proposed and investigated for voice traffic over IP networks. The CAC method is able to deal with heterogeneous voice traffic flows of ON-OFF nature with loss and delay jitter requirements. AC decisions in each module are taken as follows. If a new voice flow does not make the rate overload probability increase beyond a predefined threshold, it is admitted by the first module and passed on to the second module. The AC decision of the second module is based on criteria related to delay and loss metrics. The loss metric is the buffer overflow probability. For the delay metric, either the maximum queueing delay, or the mean queueing delay, or a given percentile of the queueing delay, is chosen as the QoS criterion. The delay percentile should be adopted as a suitable criterion because it is more meaningful in the context of statistical end-to-end jitter guarantees.

One concrete realization of the proposed approach is studied, where the REM scheme is ensured by the well-known Chernoff bound based effective bandwidth concept. The SiRM model is obtained by exploiting the recently proposed Negligible Jitter (NJ) conjecture [17]. The application of the NJ conjecture coupled with the non-preemptive scheduling scheme (with high priority service for voice traffic and low priority for best effort traffic) leads to the application of an M/D/1/K queue with exhaustive service and multiple vacations. The server vacations correspond to situations where the output link is occupied by best effort traffic. Further technical explanations on the performed queueing analysis can be found in Section 2.7 of the Queueing Models Chapter. The obtained results show that loss and delay metrics computed with the theoretical model are quite close to those obtained with simulations, particularly when the mean packet size is used in the reference SiRM model rather than the Maximum Transfer Unit (MTU) packet size as suggested previously by the original NJ conjecture. A potential extension of the proposed CAC scheme to a network is to investigate its application to each network node, together with ReSerVation Protocol-like (RSVP) signaling for a network-wide CAC decision.

Measurement-Based AC

In general, TDAC schemes aim at guaranteeing to the submitted traffic streams a QoS level in accordance to their declared profiles. Their main weakness is the dependence of their efficiency on the suitability of the a priori declared traffic parameters. As a matter of fact, since tight values are hardly known beforehand, these parameters are usually over-estimated. Therefore, the authors of [18] propose a new method for more precise evaluation of the traffic descriptors. A recognized approach for traffic characterization is based on the token bucket mechanism which is described by two parameters: the token accumulating rate and the bucket size. Note that these parameters constitute a base for effective bandwidth evaluation and they determine the amount of resources the network should dedicate for handling the corresponding connection. The proposed method is based on online traffic monitoring, which allows to obtain the minimum effective bandwidth and to reduce the required resources. The approach is investigated for handling non real-time VBR traffic with zero packet loss as a QoS requirement. In this case the effective bandwidth is calculated according to the Elwalid-Mitra-Wentworth (EMW) formula [19]. The proposed algorithm for assessing the token bucket parameters is based on simultaneous traffic monitoring by a number of modified leaky bucket mechanisms. The effective bandwidth calculation takes into account the optimal point from the burstiness curve, which minimizes the EMW formula. The effectiveness of the proposed approach is illustrated by simulations for different traffic traces of Moving Pictures Expert Group (MPEG) video and Local Area Network (LAN) traffic.

Impact of Real-Time Protocols on AC

AC results are also used for the investigation of protocol design alternatives [20]. Low bit rate real-time traffic usually comes in small packets, thus leading to an inefficient ratio of Real Time Protocol (RTP)/UDP/IP header over payload size. Such a traffic is produced by voice, video and circuit switched applications, e.g., in the terrestrial radio access networks of wireless communication systems like Global System for Mobile (GSM) or Universal Mobile

Telecommunications System (UMTS). Base stations are connected to radio network controllers over low-bandwidth links because they usually produce only a small amount of traffic. As the fixed network capacity is expensive on leased lines, too, header compression techniques are attractive to save bandwidth. The assumed QoS requirements for low bit rate real-time data traffic are small packet loss and delay. The notion of Delay Budget (DB) is introduced, and the performance requirement is that the probability of packet waiting time exceeding DB remains smaller than a target value. A suitable AC mechanism for that scenario is based on the N.D/D/1 queueing system, from which the packet waiting time distribution can be derived. This model requires homogeneous flows, i.e., flows with identical, constant, packet inter-arrival times and identical, constant, packet sizes. In addition, reasonably designed buffers can prevent packet loss. The analytical results in [20] compare the critical net load for voice traffic with and without header compression. The impact of header compression on the critical load is two-fold: both the mean bit rate and the burstiness of flows are reduced. As a consequence, up to 150% more traffic can be carried over the same low-bandwidth links.

1.3.2 Admission Control for Elastic Traffic

AC methods for elastic traffic may also be based on TDAC or MBAC. The first two algorithms hereafter described fall into the first category. They use the notion of a traffic contract between the users and the network and require explicit resource reservation based on declared traffic descriptors. Contrarily to these, the third and fourth approaches below propose an implicit identification of flows "on the fly", the last one relying more specifically on a pure MBAC algorithm to design the AC scheme.

Traffic Descriptors-Based AC

The AC algorithm proposed in [5] was developed in the framework of the AQUILA project. The algorithm is designed to fit within a DiffServ architecture and to meet the QoS requirements of a network service that consists of long-lived greedy TCP connections, such as those generated by File Transfer Protocol (FTP) users. Prior to establishing a TCP connection, the application sends a request to the AC agent which is located at the edge router, specifying a target rate. This approach relies on the notion of a traffic contract that needs to be enforced by the edge routers. Each admitted TCP connection is subject to policing, based on a token bucket mechanism whose parameters are inferred from the target rate and Round Trip Time (RTT) statistics (minimum and average if available). The token bucket parameters, i.e., the token filling rate and bucket size, constitute the input parameters for admission decision. The role of the AC agent is to make sure that the traffic offered to the core network does not exceed the dedicated resources, bandwidth and buffer, on the corresponding edge router to core router link. More precisely, a new TCP flow is accepted if the following conditions are satisfied: the sum of the token bucket rates, respectively bucket sizes, assigned to the ongoing flows and to the incoming flow must be less than or equal to the amount of capacity, respectively buffer, dedicated to the corresponding service on the link. Details on this approach and numerical evaluation based on simulations can be found in [5].

Another AC algorithm [21] targets sporadic short-lived flows requiring a new service, so-called Premium Message Handling Service (MHS), in the context of a DiffServ network. This service is designed for successful delivery of short messages to destinations, using point-to-point TCP connections. The message delivery delay (transfer time) should not exceed a predefined target value. Handling such a traffic assumes that messages have a fixed limited size. Based on this assumption, the authors compute the maximum number of packets that can be sent in one burst. The transmission of a whole message typically completes during the TCP slow-start phase. At the packet level, a separate queue for Premium MHS packets is required in each router with dedicated amounts of bandwidth and buffering. The buffer size should be dimensioned to avoid packet loss and to satisfy the message transfer time requirement. At the flow level, the admission decision is based on the maximum number of admitted connections. The corresponding threshold is inferred from the traffic descriptors declared by each TCP connection, the dedicated buffer size and bandwidth, and allowed message transfer time. The traffic descriptors are the peak bit rate, which in this case denotes the link rate connecting the host to the edge router, and the maximum burst size. The impact of message size on the maximum number of admitted flows is also investigated: the bigger the message size, the smaller the acceptance area.

Measurement-Based AC

The AC procedure proposed in [22] applies to elastic traffic typically produced by Web flows corresponding to file downloads. New TCP connections are rejected, by packet dropping, when a maximum number of competing TCP connections is reached. TCP connections are identified on the fly by detecting TCP SYN packets. In the considered network configuration, the AC mechanism operates on a link transferring Web downloads that potentially constitutes a bottleneck. The maximum number of connections must be compatible with a target packet loss probability. Packet loss probability may indeed be considered as a relevant performance indicator, as it clearly impacts the throughput of a TCP connection [23]. The maximum number of admissible connections is derived by means of packet-level and flow-level analysis and results are compared with simulations. For a given number of ongoing connections, a packet-level model first computes an upper bound of the loss probability for a given buffer size, assuming the independence of concurrent connections and a perfect fair sharing of the bandwidth between connections. A flow-level model is then used to compute the distribution of the number of flows for a given load and to ultimately derive the blocking probability. The model considers a homogeneous flow RTT and attempts to capture the rate limitation due to large RTTs.

In contrast to the previous approaches, the AC mechanism presented in [24] applies to elastic flows that do not have clearly defined service requirements. Here, the objective of AC is to avoid the negative situation where demand overload, created in the absence of AC by an increasing number of flows in progress, results in ever smaller throughputs for each one of them. Demand overload can occur in any reasonably well dimensioned network for a variety of reasons, including link and/or router failures. The proposed AC mechanism is implicit in the sense that flows are not preceded by any explicit signalling exchange between users and the network. The users generate flows spontaneously as in the current best effort architecture, and the network elements performing AC detect them on the fly by maintaining a list of protected flows and systematically comparing the flow identity of all arriving packets with this list. The flow identity is determined from invariant fields in the IP and TCP packet headers. In IPv6, the "flow label" field could be used together with the origin and destination IP addresses. In the absence of congestion, a new flow is added to the list of protected flows and the packet is forwarded. If the link is already too heavily loaded, the first packet (or packets) of a new flow is simply discarded. The current state of congestion is determined based on measurements since flows do not declare any traffic descriptors. A new flow is accepted if the estimated available bandwidth is greater than or equal to a predetermined admission threshold. An optimal choice of the admission threshold should produce negligible blocking under normal load while maintaining sufficiently high throughput for admitted flows in case of overload [25]. The authors point out that the sought solution does not need to be highly accurate: elastic traffic is particularly tolerant to estimation errors and MBAC is inherently self-correcting. To demonstrate the feasibility and efficiency of the proposed AC mechanism, a testbed was set up and trials were carried out with real traffic. The AC experiments in [24] were performed on a 10 Mbit/s

Ethernet segment. New tools (bubble and box diagrams) were purposely developed to visualize the realized flow level performance. They clearly illustrate the phase change in perceived performance occurring as demand evolves from normal load to overload. Preliminary test results of an upgraded system on a higher speed Ethernet interface at 1 Gbit/s suggest that the solution is scalable and could be implemented without much difficulty in modern backbone routers.

1.4 Network Admission Control and Resilience

Traditionally, AC has been primarily understood to be the answer to the question: "How much traffic can be carried over a single link without violating the QoS requirements in terms of packet loss and delay on that link?" This kind of AC is referred to as LAC, as already stated. Examples of LAC are peak rate allocation, the concept of effective bandwidth and, to some extent, MBAC. For practical implementations, the traffic limitation on all network links is required. This can be done in a straightforward way by applying LAC methods on a link-by-link basis, e.g., by using resource reservation protocols like RSVP. Such an approach, however, induces information states about individual reservations inside the network if they are co-located with each router in the path of a flow. Another option is a Bandwidth Broker (BB) solution [26] that keeps all the information in a centralized database. Both approaches are problematic. The first one requires core routers to be aware of AC decisions and the second one presents a single point of failure. Therefore, a NAC scheme is desired whose entities only reside at the border routers. Finally, to achieve QoS in the Internet, an Inter-Domain AC (IDAC) is required to allow interworking between AC entities of different Autonomous Systems (AS).

1.4.1 Introduction to NAC Methods

A taxonomy is provided in [27, 28] of four NAC concepts that, although basically different, are all based on virtual capacity budgets which are administered at different locations in the network. Flows have to ask one or several of them for admission and are admitted if all of them have enough capacity. The size of these budgets must ensure that no congestion can occur on any link in the network [29].

- The Link Budget (LB) based NAC defines a capacity budget for each link, and flows have to be admitted by all LBs that are associated with their path. This is depicted in Figure 1.1(a) and corresponds to the above described link-by-link application of LAC methods, which induces reservation states in the network or a possible single point of failure, according to the implementation approach. In contrast, the following NAC methods require reservation states only at the border routers.

- The Ingress and Egress Budget (IB/EB) based NAC, represented in Figure 1.1(b), performs an AC decision only at the ingress and the egress routers of a flow independently of each other. This concept is known from the DiffServ context [30] and is implemented in the AQUILA project [2].
- The Border-to-Border (B2B) Budget (BBB) based NAC has a BBB for each B2B relationship, which may be consulted at the ingress border router, as illustrated in Figure 1.1(c). The relevant BBB for a given flow request is determined by its ingress and egress routers.
- Finally, in the Ingress and Egress Link Budget (ILB/ELB) based NAC, every ingress or egress border router holds a link specific budget for each link in the network. Flows ask for admission both at their ingress and egress routers, consulting any budgets related to the links on their paths, as shown in Figure 1.1(d). If only ILBs are used, the NAC scheme may be viewed as a local BB, having at its disposal a private share of the total network capacity, and is similar in that case to the hose model [31].

1.4.2 Performance of NAC Methods

The average resource utilization in a suitably dimensioned network is a reasonable performance measure to compare NAC methods, from a resource efficiency point of view [32]. Such a comparison has been undertaken by means of the COST-239 network shown in Figure 1.2, deployed within the COST Action 239 in order to evaluate possible solutions for an optical transmission infrastructure in Europe [33].

According to Figure 1.3, LB NAC reveals the best performance, followed by ILB/ELB NAC and BBB NAC. For small offered load, in terms of the average number of simultaneous flows, the differences are significant while for sufficiently large offered load, LB NAC, ILB/ELB NAC and BBB NAC converge to 100% potential resource efficiency. In contrast, IB/EB NAC has a significantly lower performance, since the NAC scheme cannot avoid pathological traffic patterns that cause congestion on some links and leave other links unused. In general, performance depends on many aspects such as network topology [34], traffic matrix, and routing [35].

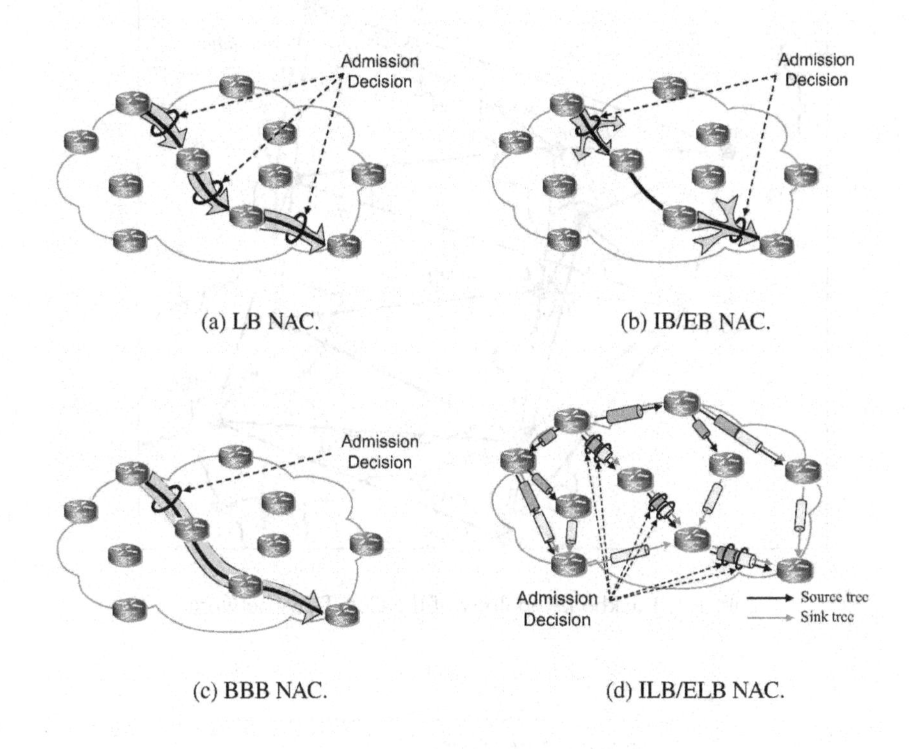

(a) LB NAC.

(b) IB/EB NAC.

(c) BBB NAC.

(d) ILB/ELB NAC.

Figure 1.1: Budget based Network Admission Control (NAC) methods

1.4.3 Capacity Renegotiation

In case of LB NAC, the LBs can be used by any flow in the network traversing the respective link, while the other budget types can be used only by a subset of flows, e.g., by those entering the network at a common ingress router. Flow blocking occurs if the budget capacity for such a subset of flows is exhausted. This can happen while the network is still far from being overloaded, e.g., if other budgets are hardly used. In such cases the network capacity is badly distributed among the budgets and capacity should be renegotiated. The AQUILA architecture is conceived to perform such a renegotiation in a layered and scalable way [36], and algorithms to achieve the redistribution are investigated in [37]. Capacity renegotiation among border routers needs some signalling efforts and should not be triggered for very small capacity units. Hence, capacity budgets are increased by a chunk of bandwidth, which leads to over-reservation for traffic aggregates. The tradeoff between over-reservation and signalling reduction is investigated in [38], where analytical results are presented.

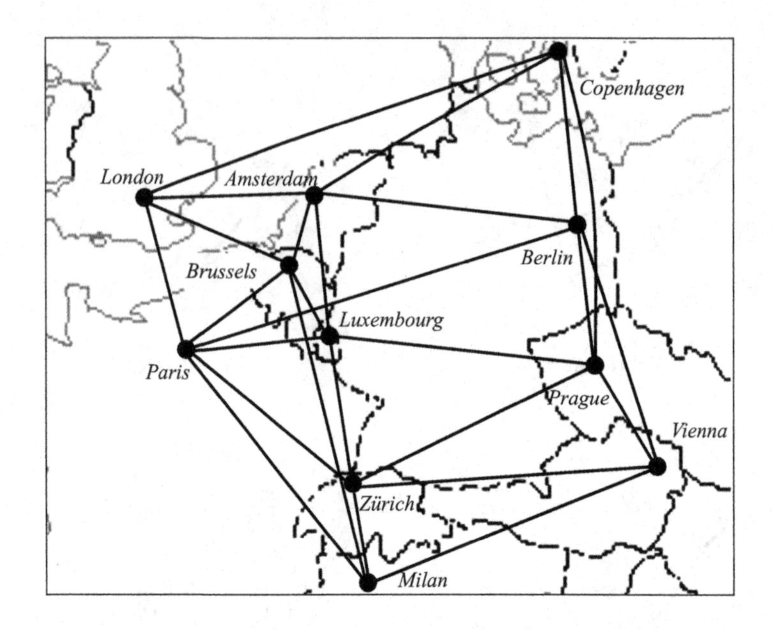

Figure 1.2: Backbone topology of the COST-239 network

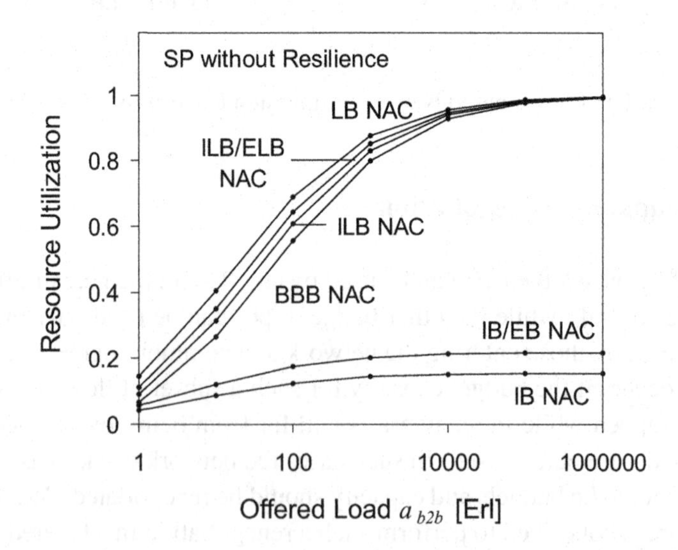

Figure 1.3: Average resource utilization without resilience requirements for different NAC methods depending on the offered B2B load in the COST-239 network

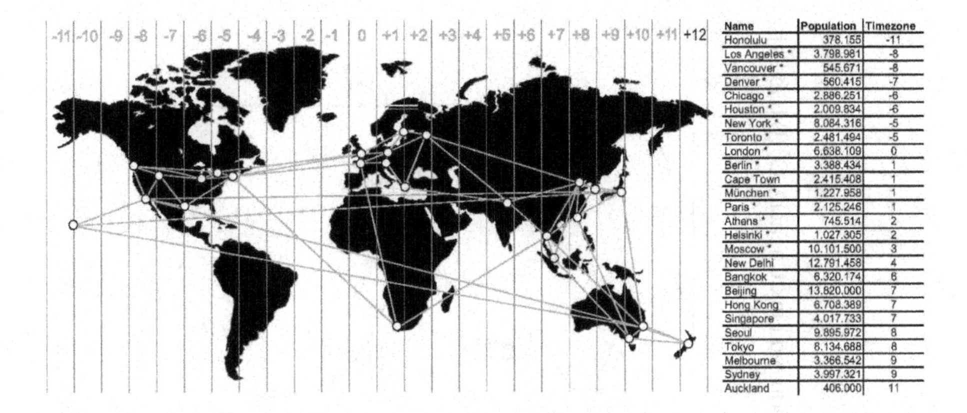

Name	Population	Timezone
Honolulu	378.155	-11
Los Angeles *	3.798.981	-8
Vancouver *	545.671	-8
Denver *	560.415	-7
Chicago *	2.886.251	-6
Houston *	2.009.834	-6
New York *	8.084.316	-5
Toronto *	2.481.494	-5
London *	6.638.109	0
Berlin *	3.388.434	1
Cape Town	2.415.408	1
München *	1.227.958	1
Paris *	2.125.246	1
Athens *	745.514	2
Helsinki *	1.027.305	2
Moscow *	10.101.500	3
New Delhi	12.791.458	4
Bangkok	6.320.174	6
Beijing	13.820.000	7
Hong Kong	6.708.389	7
Singapore	4.017.733	7
Seoul	9.895.972	8
Tokyo	8.134.688	8
Melbourne	3.366.542	9
Sydney	3.997.321	9
Auckland	406.000	11

Figure 1.4: A wide area network with B2B traffic aggregates exhibiting busy hours at different times

BBBs for NAC as well as Label Switched Paths (LSPs), used for traffic engineering in Multi-Protocol Label Switching (MPLS) technology, can be viewed as capacity tunnels. If such capacity tunnels are used, bandwidth can be allocated to them either statically or dynamically. The traffic rates of B2B traffic aggregates usually fluctuate on the basis of a 24 hour period. Therefore, capacity renegotiation or Multi-Hour Design (MHD) is especially useful to save some amount of capacity in wide area networks where B2B aggregates sharing a common link may show different phases. In [39], the capacity savings of Adaptive Bandwidth Allocation (ABA) vs. Static Bandwidth Allocation (SBA) are investigated. Dynamic traffic matrices are built in such a way that the B2B traffic aggregates scale with the activity of their ingress and/or egress routers. Depending on these traffic models, from 2% to 18% of the capacity can be saved in the network of Figure 1.4 when the B2B aggregates of the traffic matrix are proportional to the city sizes. Figure 1.5 depicts the obtained time-dependent capacity requirements for the link from Seoul to Tokyo. Of course, SBA always requires the busy hour capacities for all carried B2B aggregates, while the needed capacity for ABA fluctuates on a significantly lower level. Consequently, nearly 50% of the capacity can be saved with ABA on that link when considering Linear to Provider and Consumer Activity (LPCA) traffic model, while almost none can be saved with Linear to Consumer Activity (LCA) traffic model, see [39]. On other links, hardly any capacity can be saved with either traffic model. In addition to the traffic model, the potential for capacity savings is also influenced by routing, responsible for traffic composition on the different links. Maximization of the capacity sav-

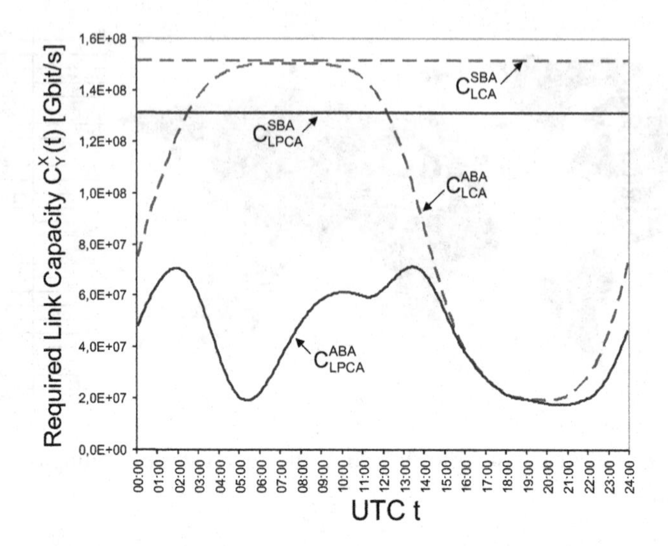

Figure 1.5: Time-dependent capacity requirements on the Seoul - Tokyo link with SBA and ABA, for the LCA and LPCA traffic models

ings is obtained when B2B aggregates with non-overlapping busy hours are transported, as much as possible, on the same links.

1.4.4 Performance of NAC Methods with Resilience

An important feature for NGNs is the provisioning of reliable QoS services, i.e., QoS should not be compromised in case of short-term local network failures [1]. If a local outage occurs, traffic is quickly re-routed around the failed network element and is still delivered to its destination. QoS can be maintained only if sufficient capacity is available on the alternate routes. In this case the network is resilient to failures. For this purpose, a set of potential failure patterns, e.g., all single link failures, together with the corresponding re-routing schemes, has to be taken into account when the capacities of NAC budgets are configured. The performance of NAC methods is investigated under these side conditions in [28, 40]. According to Figure 1.6, BBB NAC shows best performance in terms of resource utilization under resilience requirements, followed by ILB/ELB NAC and LB NAC. IB/EB NAC is again the least efficient method. Bandwidth utilization also increases for all NAC methods with increasing offered load, but clearly remains below 70% even in the best case. For very large offered load, the reciprocal of that value is the ratio of the total capacity, normal plus backup, required in the failure scenario to the capacity required in normal situation. The mere backup capacity is in

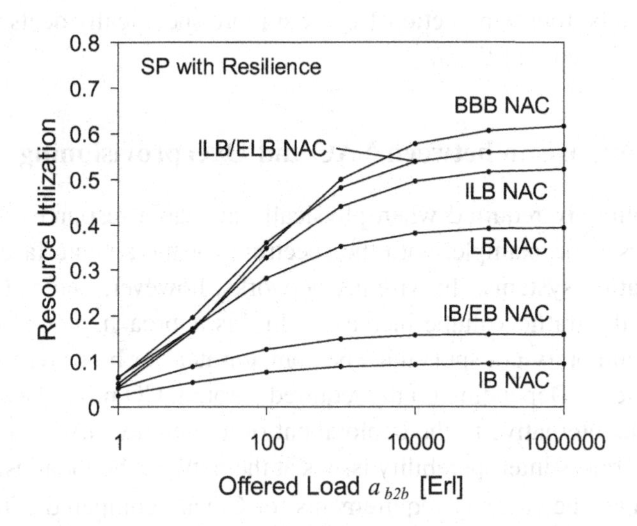

Figure 1.6: Average resource utilization with resilience requirements for different NAC methods depending on the offered B2B load in the COST-239 network

the range from 50% to more than 100%. As BBB NAC shows the best performance under resilience requirements, it is therefore implemented in the KING testbed.

The results in [40] also show that routing has a definite impact on the required backup capacity, particularly in networks with a high offered traffic load. A routing optimization problem under resilience requirements exists if backup resources can be shared by various flow aggregates in different failure scenarios. In this case, the required backup capacity is a performance measure which should be minimized. In situations of high offered load, the impact of traffic dynamics vanishes and the routing optimization problem can be solved for static aggregate sizes as given by the traffic matrix. In [41] the concepts of Path Protection (PP) and Self-Protecting Multi-path (SPM) are suggested. They are both B2B protection mechanisms and can be implemented by means of LSPs in MPLS. An optimization algorithm for the layout of these structures and for multi-path load balancing is presented. The results show that for some networks, when using the SPM approach, only 17% backup capacity is required to protect them against all single link and router failures. The amount of required backup capacity mostly depends on the network topology: strongly meshed networks favours the existence of link- and node-disjoint multi-paths and thus contribute to derive cost-efficient solutions. Further details on this proposal can be found in Section 1.8.3 that more specifically deals with traffic engineering based on MPLS.

1.4.5 Comparison between NAC and Overprovisioning

NAC is definitely required when physically available resources are limited, as is the case, for example, with the spectrum for the air interface of mobile communication systems. In wireline networks, however, one can easily add capacity if the traffic volume increases. In this approach, stochastic fluctuations and temporary hot spots must be compensated for by provisioning some excess capacity. Depending on the required amount, CO might be an economically viable alternative to the deployment of a complex NAC system which, in addition, raises interoperability issues at the network boundaries.

Therefore, the capacity requirements for CO are compared with those for BBB NAC and LB NAC in [42] and [43], respectively. The KING testbed has been dimensioned such that either a flow blocking probability of 10^{-3} is achieved for AC, or a QoS violation probability, defined as the fraction of time when the traffic rate exceeds the link resources, of 10^{-6} is obtained for CO. The plots in Figure 1.7 show the required network capacity for CO, LB NAC, and BBB NAC, as a function of the offered load. The relative capacity requirements CO/AC illustrate that a network operated with BBB NAC requires more capacity than a network with CO, which in turn requires more capacity than one with LB NAC. Asymptotically, all of them require about the same amount of resources since traffic fluctuations are extremely weak for very high offered load. This is due to the underlying Poisson traffic model which is not intended to account for temporary traffic shifts nor overload situations. Such traffic events may be conveniently captured by so-called hot spots: the overall B2B traffic volume remains constant but the hot spot node generates and attracts f_h more traffic than usual while the traffic volume among other nodes is slightly decreased.

With AC, network capacity is dimensioned for the nominal scenario since overload can be met by flow blocking when a hot spot occurs. The increased blocking probability created by hot spot occurrence is illustrated in Figure 1.8. With CO, capacity is dimensioned so that it is able to carry the whole traffic with the desired maximum QoS violation probability of $p_v = 10^{-6}$, for all possible single hot spots. As a consequence, the required capacity increases with the hot spot factor f_h. Figure 1.8 also shows the capacity requirements for CO relatively to BBB NAC and LB NAC, as a function of the hot spot factor. Now, the network operated with CO needs more resources than operated with AC methods, for sufficiently high hot spot factors. It is remarkable that the

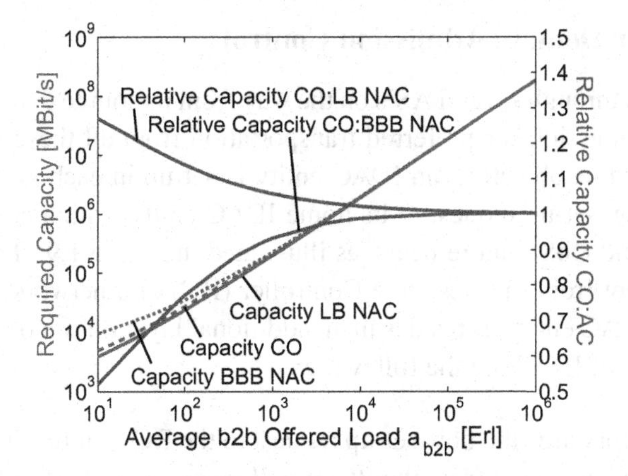

Figure 1.7: Required capacity for CO, LB NAC and BBB NAC, and the corresponding ratios

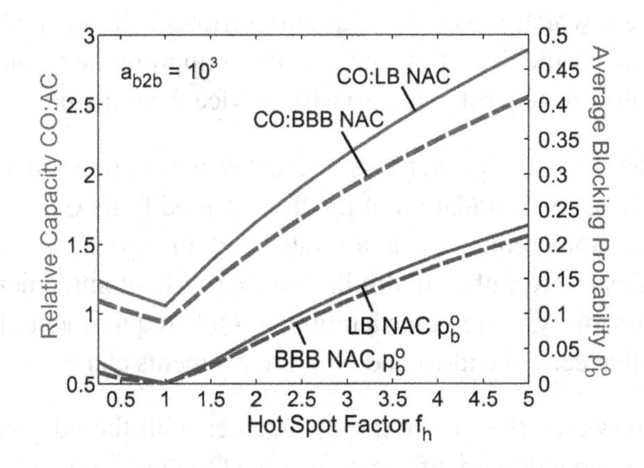

Figure 1.8: Required capacity for CO relative to LB NAC and BBB NAC, and the corresponding blocking probabilities for both NAC methods

additional capacity does not scale with the hot spot factor. The reason for that is the following. A link carries aggregated flows leading to or coming from a hot spot, therefore their rates scale with f_h. But, in addition, the links also carry transit traffic whose rates are possibly slightly decreased by the hot spot event (recall that the overall traffic volume is kept constant). Hence, the required capacity of a link depends on its traffic composition, i.e., on the traffic matrix, routing and network topology. As a consequence, the described analysis can be appropriately applied to dimension networks operated with CO to make them safe against certain hot spot events.

1.4.6 Inter-Domain Admission Control

If a flow goes through several AS's on the way from its source to its destination, it requests admission for preferred transportation from all these intermediate systems. For this objective, an IDAC entity is set up in each system. A user submits a flow set-up request to its home IDAC entity, which contains traffic descriptors and QoS requirements, as illustrated in Figure 1.9. The IDAC entity is different from the Resource Controller (RC) of a network, which is an intra-domain AC entity; it resides in an additional layer on top of the RCs.

The tasks of IDAC are the following:

- IDAC forwards the flow set-up request to the RC. If it receives a positive acknowledgement from the RC, it will go on to the next tasks.
- The packets of the flow suffer a variable delay due to multiplexing with cross-traffic in the network. Therefore, the inter-packet distance may be shortened, which affects the traffic descriptors. Thus, IDAC evaluates the traffic profile distortion and calculates appropriate traffic descriptors for the flow to request some priority service from the next domain.
- IDAC assesses the packet loss and delay within its domain and determines the QoS degradation of the flow to modify its QoS requirements. Note that some components are rather additive, e.g., the packet delay, while others are rather multiplicative, e.g., the complementary packet loss probability. The consideration of QoS requirements is needed to ensure that certain end-to-end QoS requirements can be met.
- IDAC forwards the flow request, together with the adapted traffic descriptors and modified information about QoS requirements, to the IDAC entity of the following domain. This task may require some information about inter-domain routing.

This scheme is repeated in each intermediate system until the flow request reaches the destination domain. A flow request can be blocked by any of the IDAC entities of the intermediate systems if their available resources are not large enough to fulfill the end-to-end QoS requirements of the flow.

The distortion of traffic descriptors is studied in [44]. The authors assume token bucket parameters to characterize the flow profiles. They propose to assess the flow distortion by considering the quantile-based IP Packet Delay Variation (IPDV), which is measured between edge routers [45]. Therefore, the minimum IP packet transfer delay $IPTD_{min}$ and the $1 - \epsilon$ quantile of the IP packet transfer delay $IPTD_{upper}$ are estimated [46]. The packet jitter, i.e., the

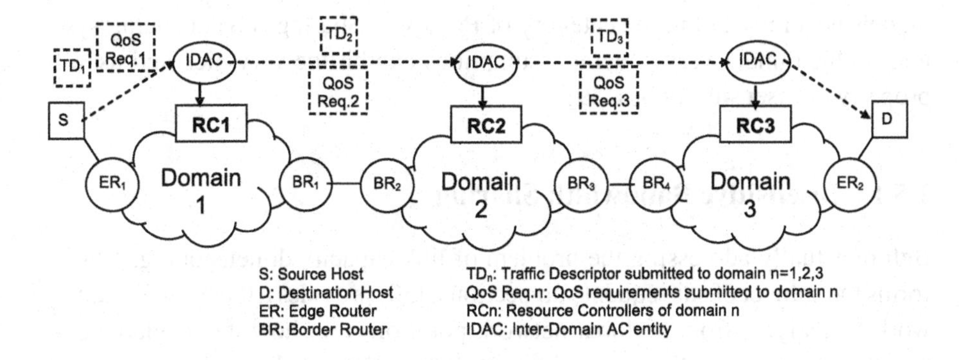

Figure 1.9: Inter-Domain Admission Control (IDAC)

delay variation, is next computed as $IPDV = IPTD_{upper} - IPTD_{min}$. The bucket size b_{next} for the next domain is then calculated by $b_{next} = b + IPDV$, b being the signalled bucket size for the flow in the current domain. The paper investigates the accuracy of this estimation method by means of simulations and finds that it provides a conservative upper bound for the bucket parameter of the flow that leaves the system.

1.5 Capacity Dimensioning

To correctly dimension network links it is essential to understand the three-way relation between link capacity, expressed demand, and realized quality of service. It is well known that Internet traffic is very complex at the packet level: self-similarity and Long Range Dependence (LRD) are commonly reported phenomena (see the Traffic Measurement, Characterization, and Modeling Chapter, Sections 3.3 and 3.4). Such a relation is thus very difficult to establish in this domain, so it is very appealing to develop performance models at a higher level, namely that of traffic "flows" or "sessions." One can refer to [47], for example, to get a thorough discussion on the concept of a flow and some possible definitions. For elastic (data) traffic, which still forms the vast majority of the Internet traffic today, it is appropriate to express dimensioning objectives in terms of the mean flow throughput, or equivalently the mean transfer time as perceived by users to download digital documents. Assuming a Poisson flow arrival process, corresponding to an infinite-size source population, and fair sharing of the available bandwidth by concurrent flows, supposed to be an ideal achievement of TCP dynamic control, the performance models generally fall in the category of Processor Sharing (PS) queueing systems, which under very general assumptions exhibit the very nice and useful property of insensitivity [48, 49].

1.5.1 Insensitive Bandwidth Sharing

Before actually addressing the problem of link capacity dimensioning, which forms the subject of subsequent sub-sections, [50] provides a theoretical framework to analyze, from a global network point of view, how the available capacity is allocated to flows in a data network, or how it should be in order to preserve some interesting properties such as fairness or performance insensitivity [51]. The used generic flow model represents the data network as a set of links where a random number of flows compete for access to the link bandwidths. Flows may pertain to different classes, each of which characterized by their route, average traffic demand, and possible per-flow rate limitations, such as access rate constraints. The approach here aims at studying flow-level dynamics and so is in contrast to some well-known models in the literature based on the notion of utility and relying on static scenarios of bandwidth allocation among a fixed number of flows. It is first shown that a PS queueing network with state-dependent service capacities can represent the flow-level model of data network with any traffic characteristics: flows are allowed to be generated within sessions, composed of successions of flow transfers and "think times"; moreover, flow sizes and think times may have very general distributions and arbitrary correlation structures. Then, using key properties of Whittle queueing networks [52] and only assuming a Poisson session arrival process, it is shown that the class of allocations which are insensitive to the detailed traffic characteristics, except for the traffic intensity on each route, are those for which the balance property holds. The balance property expresses the fact that the relative change in the bandwidth allocation to class j, when a flow of class k is removed, is symmetric in j and k.

In such conditions, the stationary distribution of the number of flows process writes down as a function of the traffic intensities and allocation functions only. It is further shown that the mean duration of a flow is proportional to its size for each class; thus, an expression of the mean throughput is easily obtained. Among all possible insensitive allocations, the most effective one, i.e., the unique one which saturates the capacity constraints, is identified as *balanced fairness* [53], for which a recursive expression is provided that allows its numerical evaluation. In [50], these results are applied to some simple specific network topologies, such as trees and hypercubes, where the balance function has a closed-form expression. It is shown in particular that utility-based allocations, including Max-Min fairness as a limit case, do not satisfy the balance property, and thus are not insensitive, except for proportional fairness in the case of homogeneous hypercube topologies.

1.5.2 Access Network Links

For access networks, a typical example of the above cited performance relationship is the Engset loss formula for classical circuit-switched telephony. In [54] it is shown that similar relations exist for high speed IP access networks carrying data traffic. Assuming a finite-size source population and fair sharing of the capacity among the user-generated concurrent flows, some generalized Engset formulas with PS discipline are derived, relating capacity, demand expressed in terms of the per-user offered traffic, and performance quantified by the useful per-flow throughput. Performance is shown to be largely independent of precise traffic characteristics such as the statistical distributions of flow size and think time, as recognized in [55].

Very simple approximations of this exact model can be obtained, exhibiting two distinct performance regimes and thus leading to two corresponding strategies for dimensioning: a transparent regime, where the mean useful rate equals the access rate, and a saturated regime, where the whole capacity is always used and fairly shared by active users. Analysis of the impact of a possible heterogeneous demand demonstrates that a homogeneous demand constitutes a worst case, so that considering an average offered traffic would be a conservative strategy for dimensioning. From these results, the definition of simple dimensioning rules follows and is illustrated in Figure 1.10. Note that operation in the saturated regime, in order to meet a target useful rate lower than the access rate, might be quite risky since the performance is highly sensitive to the accuracy of the offered traffic estimate.

Extensions to the approximate model are also considered in [54], to handle situations with unfair bandwidth sharing or different access rates. Broadly speaking, it remains that the key parameter for dimensioning is the offered traffic, defined as the average data rate a user would generate in the absence of congestion.

This kind of access model, i.e., handling source populations of finite size, is generalized in [56] by considering possible state dependent blocking probabilities and capacities. Such model extensions are intended to tackle flow level performance when the network transports both streaming and elastic traffic, with a fluctuating available capacity for elastic flows, or when full acceptation of the incoming flows cannot be guaranteed due to transmission layer capabilities, such as in some wireless access networks like Wideband Code Division

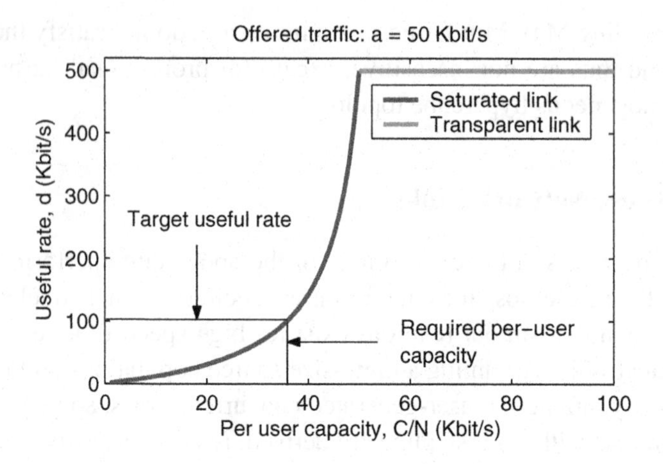

Figure 1.10: Dimensioning function based on approximations - Access rate = 500 kbit/s

Multiple Access (WCDMA). Some new insensitivity results of PS network models are obtained in this context, and the main performance parameters are derived, including state probabilities, blocking probabilities and mean sojourn times of elastic flows. Further technical description of this contribution can be found in the Queueing Models Chapter, Section 2.5.

1.5.3 Backbone Links

Best Effort Provisioning

For dimensioning IP backbone networks, one possible approach is bandwidth (over-)provisioning, where network transparency is ensured without implementing any specific QoS mechanism. Here "transparency" means that traffic flows are not constrained by the considered network links and can make full use of their access rates. In line with this approach, [57] suggests to make use of standard traffic measurements performed by network operators on a relatively large time scale, e.g., 5 or 15 min, while the time scale at which the QoS is perceived in terms of transparency is much finer, about 1 s. In this framework, the adequate QoS indicator is defined as the probability, or fraction of time, that the aggregated rate of offered traffic is greater than the link rate.

Two main assumptions seem appropriate for backbone links with high level of traffic aggregation and high capacity with respect to common access rates: (i) the aggregated traffic intensity is Gaussian; (ii) the number of concurrent flows results from a fluid M/G/∞ model [58]. Under these assumptions, the-

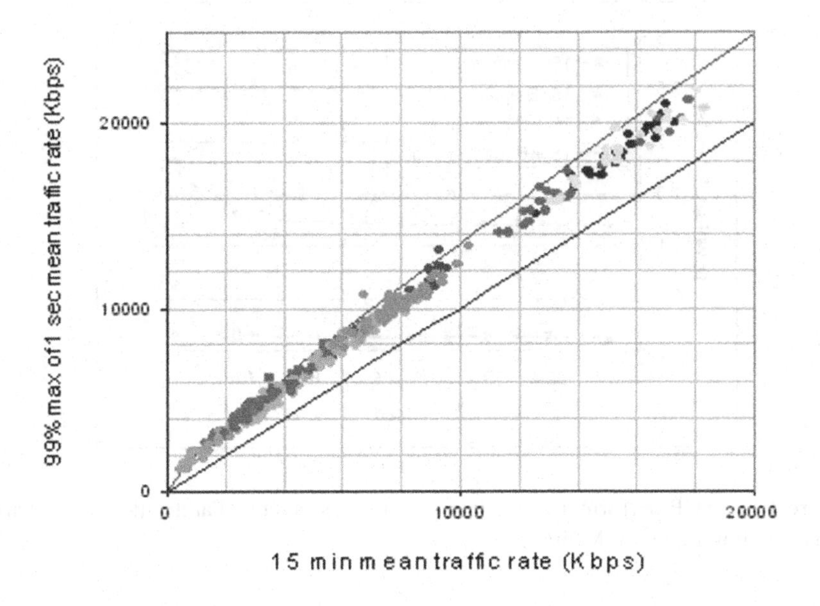

Figure 1.11: Peak rate vs. long term average rate on a backbone link with ADSL traffic

oretical expressions of the required capacity are derived in [57] as a function of the average offered load and the transparency time scale. A simple explicit formula can be obtained in case of exponential flow size distribution. An experimental validation of the method is performed in an operational network environment with Asymmetric Digital Subscriber Line (ADSL) traffic: as can be seen in Figure 1.11, the model is shown to provide a very good fit to the experimental data relating the long term average rate to the "peak rate" expressed as the 99% quantile.

QoS Differentiation

Another approach to dimension TCP/IP networks is to support QoS differentiation. The performance of Internet links carrying TCP traffic with two priority classes is considered in [59, 60]. The elastic flows are assumed to appear according to a Poisson process and to fairly share the capacity (link bandwidth). Hence the stochastic process of TCP flows is described by a multi-server PS queueing model, M/G/S-PS, with two customer classes. The "number of servers" S of the queueing system is the ratio of the link capacity to the access rate, assumed identical for all users.

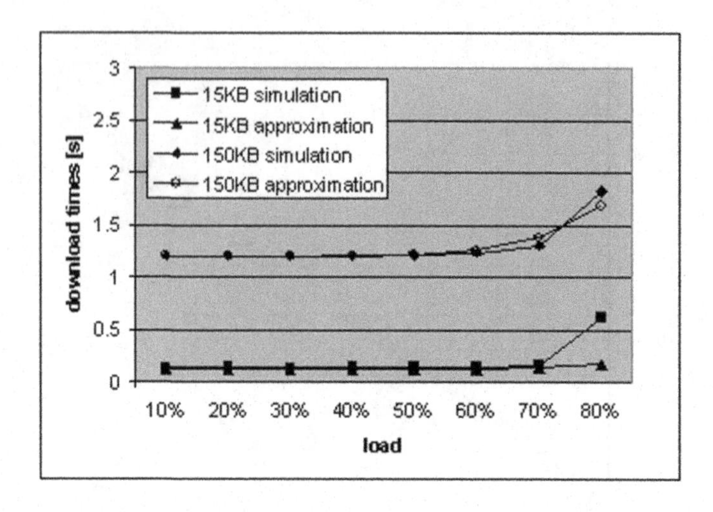

Figure 1.12: TCP performance for low priority customers - Pareto file size distributions, and link rate = 10 Mbit/s

Considering that high priority flows have strict priority with preemptive resume over low priority flows, closed-form expressions are derived for the mean transfer time of high priority flows in the general case, based on known results for the Generalized Processor Sharing (GPS) queues [48], and of low priority flows in some special cases with exponential flow size distribution [61]. In the general case, only approximate expressions for the mean transfer time of low priority flows can be provided. By comparing with `ns` simulations, it is demonstrated that the model provides highly accurate performance predictions, as shown in Figure 1.12, when the mean flow size is at least 10 IP packets, the loss rate is negligible, and the total traffic load does not exceed 70–80%. When those conditions are not met, some packet level TCP features such as slow-start and packet loss effects create significant discrepancies between simulation results and the simplified flow model outputs.

1.5.4 Streaming and Elastic Traffic Integration

TCP is still the dominant transport protocol in the Internet: according to recent experiments, it generates more than 90% of the overall traffic measured in bytes. Nevertheless, it is not possible to ignore the current quick development of real-time applications generated by voice and video services. So, in the near future, streaming and elastic traffic will coexist in the Internet, each of them with a non-negligible proportion of the overall traffic, and the problem of their integration within shared resources has to be considered. Two contributions in COST 279 have provided substantial advances in that direction, by investigating the performance of TCP data transfers in presence of a time-varying bandwidth. This variable capacity is that which is left available by streaming flows, the packets of which are supposed to have priority with respect to those of elastic flows.

Flow Level Dynamics

First, [62] concentrates on flow level TCP performance by considering a dynamic system where elastic flows as well as streaming flows appear and disappear according to given stochastic processes. Assuming Poisson flow arrivals on a network link and fair bandwidth sharing among elastic flows, the model is based on the performance of an M/G/1-PS queue with time-varying capacity. It is shown that the mean transfer time, or equivalently the mean throughput, of elastic flows may become greatly unstable if the streaming traffic volume is not controlled. On the contrary, when the conditions for uniform stability are met, e.g., by performing admission control on the streaming calls, the available bandwidth is always greater than the elastic traffic demand so that the realized throughput is good and more easily predictable.

In the case of uniform stability, stochastic bounds for the mean transfer time are given which are shown to be rather tight and to depend only on the mean streaming and elastic loads, being insensitive to other detailed traffic characteristics. These bounds correspond to two extreme regimes defined by the time scales at which the streaming flow and the elastic flow processes evolve. The fluid regime is a best case providing the lower bound for response time, where the streaming flow process evolves so fast that elastic traffic sees a constant available bandwidth. At the opposite, the quasi-stationary regime is a worst case providing the upper bound, where elastic flows see a succession of equilibrium states with a constant number of streaming flows. A set of simulation experiments illustrates how the elastic traffic performance evolves between these two bounds, as a function of elastic load and under different scenarios for the streaming and elastic flow size distributions. For example, Figure 1.13 shows, rather surprisingly, that in the local instability region the elastic flow throughput is higher when the variability of flow size is high; besides, system insensitivity is verified in the uniform stability region.

TCP Behavior

As a complementary approach, [63] deals with packet level dynamics of TCP connections: an analytical model is developed which is able to reproduce the

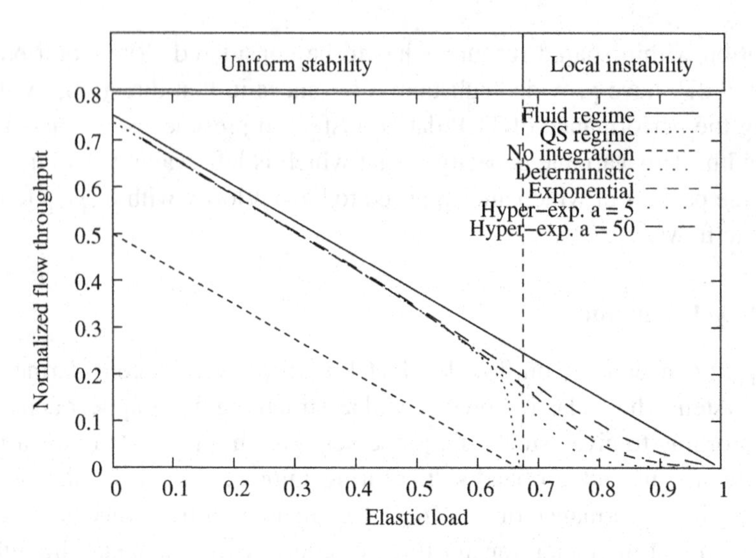

Figure 1.13: Impact of the elastic flow size distribution on the performance (elastic flow throughput) of streaming/elastic traffic integration

main features of TCP (Reno) behavior in the presence of time-varying available capacity. The interaction between one persistent TCP flow and high priority real-time traffic, e.g., voice and video, is modeled by a two-dimensional continuous time Markov chain where the state vector is defined by the number of current streaming flows and the size of the TCP congestion window.

The accuracy of the proposed model, as compared to extensive `ns` simulations, is shown to be satisfactory on the whole, and especially good when the buffer size is of the same order as the bandwidth-delay product of the connection and when the time scale of streaming flow dynamics is much larger (e.g., one order of magnitude) than the RTT. Besides, as shown in Figure 1.14, the model provides better results than other popular TCP models which appear either optimistic [64] or pessimistic [23] in predicting the average flow throughput. Finally, among other interesting TCP features derived from the model and simulations, an important result which confirms earlier simulations is the relative inability of TCP to make full use of the available bandwidth when sharing the capacity with higher priority traffic: an efficiency coefficient of about 70–80% is typically observed with respect to the ideal case of constant capacity and correctly sized buffers.

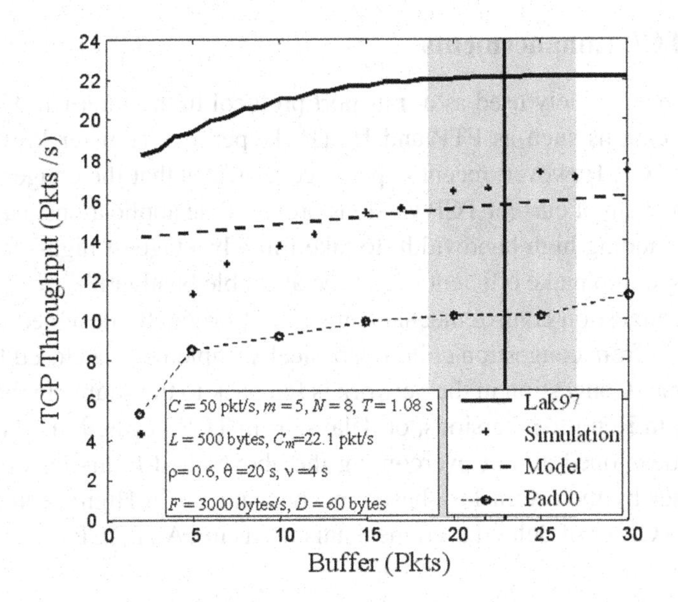

Figure 1.14: TCP connection performance from models and simulations - Streaming traffic (VoIP) offered load = 0.6

1.6 TCP and Packet Level Performance

Although not initially designed to do so, the family of TCP/IP protocols is widely used today, and will with little doubt continue to be used in the near future, to provide support for ultra high-speed communication technologies over a multimedia Internet network, covering the fields of both computer communications and traditional communications. To complement the flow level approach for performance analysis, as assumed in previous sections, one may believe that proper adaptation of TCP/IP to this new telecommunications world and its requirements should be necessary and could be achieved in combination with QoS and connection control mechanisms, like DiffServ and RSVP.

In [65] is made a detailed study, based on OPNET simulations of Internet services such as Web and e-mail, of the impact of the link layer, assumed Frame Relay, and TCP control parameters on packet delays and application response times. From this work, guidelines are obtained about resource dimensioning for the support of Service Level Agreements (SLAs) in IP networks. A particularly interesting point is that the end-to-end packet delay along the Frame Relay Permanent Virtual Circuit (PVC) path mostly results from the queueing delay at the Wide Area Network (WAN) ingress points, due to the significant bandwidth differences between LAN and WAN link layers.

1.6.1 TCP Enhancements

TCP has been widely used as a transport protocol in the Internet. Many popular applications such as FTP and HTTP (HyperText Transfer Protocol) are based on TCP. However, recent experience indicates that the congestion control mechanism of current TCP versions prevents the applications, particularly those demanding high bandwidth, to take full advantage of high-speed wide-area lines and to make efficient use of the available bandwidth.

The congestion control mechanism of TCP, originally designed to protect the Internet from congestion collapse, basically implements a closed loop control scheme. Congestion in the network is fed back to the source in the form of losses, as in Reno-like versions, or delays, as in TCP Vegas. Versions of TCP recently developed aim at overcoming the above-cited limitations of current TCP, either by the use of loss-based mechanisms, as in HighSpeed TCP and Scalable TCP, or of delay-based mechanisms, as in FAST TCP.

Loss-Based TCP Versions

The major reason for under-utilization by traditional TCP is related to the mechanisms used for the control of the TCP congestion window size: the additive increase mechanism is too slow and the multiplicative decrease is too harsh. In a steady-state environment, with a packet loss rate p, the average congestion window size is roughly $\frac{1.2}{\sqrt{p}}$ packets. This places a serious constraint on the congestion window that can be achieved by TCP in realistic environments.

HighSpeed TCP [66] (see also, among others, [67, 68, 69]) is a recently proposed revision to the TCP congestion control mechanism, specifically designed for use in networks with high bandwidth-delay product. HighSpeed TCP is formally defined in RFC 3649, which also gives guidelines for protocol parameter values for typical operating environments on the basis of simulation results provided in [70].

In [71] is presented a study on the qualitative and quantitative aspects of HighSpeed TCP performance in different environments, with special interest in apparent incompatibilities between the parameter settings used for obtaining the results in [66] and in [70]. The dynamics of the congestion window process and different performance metrics like the link utilization and fairness are analyzed. For example, a mixture of both regular TCP (REGTCP) and HighSpeed TCP (HSTCP) competing for bandwidth in ideal condition, with different link speeds, is considered. As shown in Figure 1.15, HSTCP provides better link utilization than REGTCP and, in the condition of that mixture of flows, this advantage increases with the link speed. In addition, [71] also presents a study comparing the performance of SACK TCP, TCP Reno and NewReno.

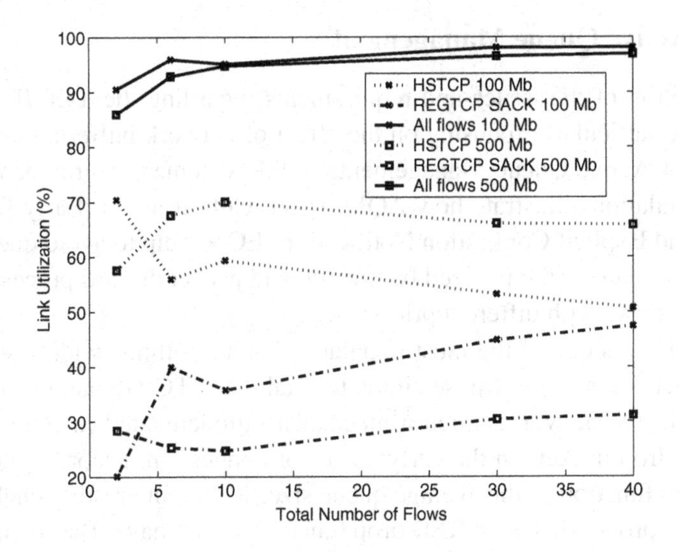

Figure 1.15: Link utilization for a mixture of REGTCP and HSTCP flows

Delay-Based TCP Versions

To envisage TCP enhancement in high speed networks, there is an important motivation to re-examine delay-based TCP algorithms like TCP Vegas: TCP Vegas has inherent pricing schemes in its design that resemble some congestion pricing schemes proposed in the literature. The better the understanding of TCP Vegas inherent pricing schemes is, the more insight into designing pricing schemes for TCP traffic we have. One promising proposal is FAST TCP [67]. Since the design of FAST TCP is strongly based on that of TCP Vegas, there is a need to reconsider both the benefits and the drawbacks of TCP Vegas in order to gain insight into the performance and deployment possibilities of FAST TCP in the future Internet.

Game theory (see [72] for a comprehensive introduction) provides us with some effective tools to analyze TCP Vegas and FAST TCP. In [73] these game-theoretic tools are used to investigate the impact of pricing schemes and TCP parameter settings on the performance of each user as well as on that of the whole network. The analysis in [73] shows that these inherent pricing schemes result in a rate control equilibrium state that is a Nash equilibrium, in game-theoretic terms, which is also a global optimum of the all-Vegas networks. It is also proved that the parameter settings of TCP Vegas and FAST TCP are much vulnerable to selfish actions from the users. Such a behavior raises serious threats on the possible deployment of FAST TCP in the future Internet.

1.6.2 Active Queue Management

A description of the issues and requirements regarding the TCP/IP protocol operation, particularly focusing on the effect of network buffering and on the analysis of Active Queue Management (AQM) schemes, is provided in [74]. Some simulations illustrate how AQM, specifically Random Early Detection (RED), and Explicit Congestion Notification (ECN) help to avoid queue overflow and to manage the realized bandwidth and packet discard process in case of various flows with different priorities.

RED [75] is one of the most popular AQM algorithms, widely studied in the Internet community, whose aim is to "stabilize" TCP dynamics in case of overload in the network. Such a mechanism implemented in router buffers begins to drop packets in the early stages of congestion, according to a drop probability function of the average queue size. It is an interesting challenge to discuss the proper shape the RED drop function should have. Based on simulation experiments, [76] shows that the standard linear drop function, considered in the original RED algorithm [75], cannot cope with a broad range of loads. Particularly, significant under-utilization of the link capacity is observed at low loads when that linear drop function is used.

Given the mathematical constraints that are derived to meet the low load and high load requirements, the authors in [76] consider two ways of determining the proper shape of the drop function, based on a class of polynomial functions and on a TCP performance model. Simulations are performed with FTP permanent connections and Web-like transfers, which demonstrate that properly derived drop functions allow a broader range of loads and yield a less varying average queue size as compared to the original linear function.

1.6.3 Quality of Voice Traffic

Today, data traffic and voice traffic are carried together over IP networks. In order to adequately support a voice service, it is necessary to understand the impact of network performance, in particular packet loss, on the voice quality. In [77] is provided an overview on how to express this impact in analytical form when quality is quantified by the Signal-to-Noise Ratio (SNR). In the first part of the contribution, it is assumed that voice samples may be lost separately and that lost samples can be reconstructed from other received samples. The methods of zero stuffing, sample repeating, and first order Linear Predictive Coding (LPC) are given as examples. In the second part, a model is presented with the more realistic assumption that groups of voice samples are lost together. Analytical results are derived when packet loss is a Bernoulli process and when lost packets are reconstructed by first order LPC. The models presented above are suitable for macro-scale voice description only. New algorithms for packet loss reconstruction are needed to provide deeper insight into the micro-scale voice structure. Thus it is necessary to study the impact of such procedures on voice quality. The SNR is used again as the measure for voice quality; its lower bounds are given in the case of individual coefficient loss and zero stuffing.

1.7 Scheduling Mechanisms

This section reports on specific works undertaken within the COST Action 279 in the field of scheduling; they are organized in three subsections dealing, respectively, with resource allocation, priority packet scheduling, and network behavior. Scheduling is considered in still a different perspective in [14], where it is shown that per-flow fair queueing is scalable in the Internet. As it is closely related to an implicit service differentiation proposal [12], this study is actually handled in Section 1.2.4 of the present Chapter.

1.7.1 Resource Allocation

A Lambda scheduler within a Generalized MPLS (GMPLS) optical node is studied in [78] by means of both OPNET simulations and an analytical model. The simulation model consists of two basic elements: the request generator and the λ-scheduler. The request generator describes the arrival process of the requests that are generated for a GMPLS node by its adjacent nodes. The user can configure the simulator so that the inter-arrival time of requests matches a desired statistical distribution. The λ-scheduler serves the requests according to the following wavelength mapping policy. First, the scheduler checks if an output wavelength is available; a constraint, i.e., a list of possible output wavelengths, can be associated to each request. Then, in order to save wavelength converters, preference is given to the output wavelength which directly maps on the input wavelength. If direct mapping is not possible, the availability of a wavelength converter is checked. If there is no free converter, the request is rejected. The developed analytical model is based on a Markov chain, and assumes the request arrival process to be Poisson and the output wavelengths to be randomly chosen. The model accuracy is validated by comparison with simulation experiments. Moreover, the scheduler performance is evaluated for different values of node size, number of converters, and offered load. Performance is evaluated in terms of node utilization and call rejection probability. It clearly degrades with increasing traffic load and improves with increasing node size and number of wavelength converters.

The authors of [79] consider a link shared by high priority traffic and some low priority streams. In order to optimize bandwidth allocation, the algorithm which decides how to share the available bandwidth between low priority flows can use estimates of the future load. The document provides a framework to evaluate the possible benefits that dynamic resource management algorithms might achieve by using traffic prediction techniques. The performance of a PS system is evaluated by simulation. In the considered experiments, the traffic prediction technique is based on adaptive filters which are able to properly react to traffic variations. Moreover, the objective of the resource allocation scheme is to try maintaining the different input queues at about the same size.

1.7.2 Priority Packet Scheduling

Bandwidth allocation to flows belonging to traffic classes with different priorities is also the topic of [80]. But, in this case, the authors consider the Modified Earliest Deadline First (MEDF) algorithm. Contrasting with other schemes that assign a fixed share to the high priority flow aggregate and that may induce QoS degradation when the number of high priority flows is larger than expected, MEDF scheduling favors high priority over low priority traffic independently of the specific traffic mix, without setting the bandwidth share in advance. MEDF supports n different traffic classes, with a First-In-First-Out (FIFO) queue being maintained for each class. Packets are time-stamped with a deadline which is a function of the packet arrival time and class priority. The main focus of [80] is on the impact of MEDF on TCP flow performance, relying on simulations to evaluate the TCP throughput for high and low priority flows. In addition, the study addresses the effects of the prioritisation level and also considers single and multiple link scenarios.

A new method called Priority Forcing Scheme (PFS) is analyzed in [81]. PFS allows Internet applications to get better service than best effort by means of the following mechanism. Besides data packets, the PFS application generates additional (redundant) packets, named reservation packets, whose role is to implicitly reserve resources for a given data flow in the network routers. Indeed, on arrival at a router interface, a data packet first checks if any reservation packet belonging to the same flow is already queued. If so, the data packet can replace the reservation packet, so that the data flow receives prioritized service. Numerical results show that performance improvements achieved by PFS increase with the average number of queued packets, i.e., as the traffic intensity increases. By properly setting the scheme parameters, packet delay characteristics can be shaped. As special cases of practical interest, VoIP and FTP applications are considered. The proposed scheme improves the end-to-end quality of both services, namely packet loss and packet delay for VoIP and TCP goodput for FTP, and is particularly efficient under overload traffic conditions.

1.7.3 Scheduling and Network Behavior

Instability phenomena in underloaded packet networks are investigated in [82]. Instability may occur in underloaded networks under particular conditions, such as routes that make the customers visit the same queues several times, variations of the customer service times at different queues, or complex scheduling algorithms. In [82], QoS schedulers in packet networks are considered. In order to represent the approaches currently considered for QoS provisioning in the Internet, e.g., DiffServ, the analyzed scenarios have acyclic routes and service times varying according to the channel capacity. Two kinds of schedulers are considered: GPS and strict priority. Networks of queues with GPS schedulers were already known to be universally stable provided that nominal flow rates are greater than or equal to actual average rates. As a new result, in [82] it is shown that instability may occur when some of the actual packet rates exceed the nominal rates, which is quite likely to happen in real systems. Indeed, a mismatch between nominal rates and effective rates is much probable since the estimation of effective rates is typically based on local traffic measurements and thus tends to be inaccurate, at least in current IP networks without CAC procedures or traffic shaping policies. Regarding strict priority schedulers, they are proved to be stable provided the priority ordering of packet flows is the same for all queues in the network. Moreover, weak forms of instability can be observed in networks which combine FIFO and priority scheduling for load values as low as 0.6.

Finally, various combinations of scheduling mechanisms and routing strategies are considered in [83] in a network of packet-switched non-geostationary satellites. The considered scheduling schemes include simple FIFO and traffic differentiation by means of PQ, Weighted Round-Robin (WRR) also accounting for packet length, Deficit Round-Robin (DRR), or Surplus Round-Robin (SRR). The scheduling schemes are applied to both basic routing procedures and Traffic Class Dependent (TCD) routing mechanisms, the latter accounting for different traffic types and looking for suitable paths based on the most appropriate optimization criterion. More details about this comparative study may be found in Section 4.5 of the Wireless Networks Chapter. Just note that, although TCD routing was shown to improve network performance for all the considered traffic classes (voice, video and best effort), no really significant differences between the various fair scheduling mechanisms were noticeable in the studied network scenarios.

1.8 Routing Optimization and Load Balancing

1.8.1 Introduction

Since the early days of IP networks, researchers have been working on problems related to the routing of traffic. The first major challenge was to keep the routes consistent and loop-free in the entire network domain even in the presence of equipment failures. In order to meet this goal, the focus of research and engineering was set on the automatic distribution of topology information throughout the network and on the automatic generation of routing tables. The result of these efforts was the intra-domain routing architecture of the Internet, which has remained practically unchanged until today. In this architecture, link weights, sometimes also called link costs, are assigned to each directed link in the domain and used for the calculation of paths. The paths between any two nodes in the domain are determined such that the sum of link weights over all path candidates is minimized. These paths are called "minimum cost paths" or "shortest paths". On top of this architecture, routing protocols have been developed which allow automatic dissemination of topology information throughout the network. These protocols have been designed to achieve network robustness and fast re-routing convergence in case of network faults. With this routing architecture, the traffic paths remain static as long as the link weights are not changed, meaning that traffic routing is not sensitive to load conditions in the network.

As the goal of robustness may be considered to have been achieved on the whole, the focus of research recently shifted to traffic engineering, i.e., performance optimization of operational networks. One of the most important goals of traffic engineering is to achieve efficient use of network resources, particularly the available bandwidth. Today, in IP networks, most traffic flows are not optimally mapped on to the available resources. The main reason for this is that in such traditional routing architectures and protocols, there is no straightforward methodology for performing traffic engineering; the traffic paths can only be implicitly impacted by changing the link weight settings in the network, which makes traffic allocation in an intuitive way almost impossible. New approaches and technologies for traffic engineering in the Internet are therefore required. In addition to traffic engineering, the optimization of IP networks with respect to resilience requirements is becoming an increasingly important research topic. Usually, the goal of such optimization efforts is to ensure non-degraded levels of service in the presence of single link or node failures with a minimum of extra capacity required for resilience.

In the framework of the COST Action 279, a number of approaches concerning research methodology and used technologies have been investigated. The individual proposals are introduced in the next subsections.

1.8.2 Traffic Engineering Based on IP Routing

Recently, many competing algorithms have been proposed which enhance the current routing algorithms and protocols with traffic engineering capabilities. These algorithms often employ multipath routing, because of the potential of the latter to achieve an even distribution of load in the network. However, the standard Equal Cost Multi-Path (ECMP) routing usually does not provide a sufficiently large number of multiple paths, as all paths are required to have the same minimal cost. Additionally, ECMP always distributes traffic evenly among the available paths, without taking into consideration the current traffic conditions in the network.

Multipath Routing with Dynamic Variance (MRDV) is proposed in [84] as a routing protocol which improves the performance of IP routing protocols under overload conditions. The main objective is to design a simple and scalable dynamic routing algorithm compatible with the traditional IP protocols. In the MRDV scheme, whenever a node tries to send too much traffic through a path, this traffic is distributed among several additional paths. A new routing metric called "multipath with variance" is proposed, which enables the traffic toward each destination to be carried by additional paths besides those with minimal cost. The "variance" parameter basically describes the extent to which the cost of additional paths may exceed the cost of optimal paths. The central concept of MRDV is that the number of such additional paths towards a destination dynamically depends on the extent of link overload. This multipath scheme results in a congestion reduction and leads to a better use of network resources. The MRDV algorithm has been evaluated using the `ns-2` network simulator in several different scenarios. The results show the algorithm stability as well as considerable performance improvements over traditional routing schemes.

In [85], Adaptive Multi-Path Routing (AMP) is proposed as a new routing algorithm for dynamic traffic engineering. This distributed algorithm is envisioned as an extension to current intra-domain routing protocols, that operates autonomously and continuously in the network nodes without introducing any management overhead. With AMP, every node performs load measurements on its output links. In case of congestion, the node which is sending out traffic directly on the congested link will immediately react to overload by trying to put more load on available alternative paths. At the same time, this node will send so called "backpressure messages" to each of its neighbor nodes, notifying them to which extent traffic they are sending is present on the congested link. In other words, the neighbor nodes are informed about their contribution to congestion. In turn, these neighbors will also try to offload their links directed towards the congestion hotspot and will pass similar congestion indications, in the form of backpressure messages, on to their neighbor nodes. This quasi-recursive signaling mechanism of backpressure messages is the central innovation of AMP. It enables global propagation of load information, and at the same time keeps the exchange of load signaling information local, and thus very efficient with respect to the consumption of network resources. Having a large number of multiple paths in the network is a prerequisite for efficient load balancing and, therefore, it is important to also consider additional paths besides those with minimal cost. In order to use such non-minimal paths and still strictly avoid routing loops, AMP uses a simple criterion that states that any neighbor node which is closer to the destination, in terms of cost, than the current node is a viable next hop for multipath routing. Note as an immediate conclusion that routing loops cannot happen as the cost to reach the destination node always strictly decreases along every hop of the path. AMP uses the 16-bit Cyclic Redundancy Check (CRC-16) for splitting traffic into micro-flow aggregates based on the (source, destination) address pairs, and divides the traffic among multiple paths by mapping portions of the CRC-16 hash space (i.e., a range of micro-flow aggregates) to the viable next hops for each destination. Load balancing is performed such that the relative sizes of the hash space portions for each destination are dynamically adjusted in every node. The algorithm has been implemented in the `ns-2` network simulator and simulations of the AT&T-US network, composed of 27 nodes and 47 links, have been carried out using a state of the art Web traffic model. AMP performance is compared to shortest path routing and ECMP routing for a broad spectrum of loads. The results demonstrate significant performance improvements and the stability of AMP throughout the investigated simulations.

It is well known that mixed integer programming must be used in order to determine the optimal set of link weights for an IP network. This makes the optimization problem NP-complete, requiring inappropriately long computation times to find the optimal solution for large networks, and thus forcing the use of heuristic sub-optimal methods. Therefore, a clear motivation for research is the need to improve the scalability of IP routing optimization. A novel method for the optimization of large IP networks, called "divide and conquer", is presented in [86]. The basic idea is to split the original network into smaller sub-networks and then to individually optimize these sub-networks. In the "divide" phase of the algorithm, the network to be optimized is decomposed into three sub-networks. Two of them are disjoint and relatively large, the third one is designed to be a small central "link" between the former two. Once that decomposition of the original network is achieved, the "conquer" phase can begin. In this phase, mixed integer programming is used to formulate the optimization problem for the two large sub-networks previously created. In the final step of the algorithm, it remains to merge the two individual results into one solution for the entire original network. The performance of the decomposition algorithm itself and that of the overall routing optimization scheme have been evaluated. The link weights setting obtained from the overall divide and conquer algorithm has been compared both to that obtained from direct optimization of the original network and to that of the non-optimized network. The topologies of the AT&T-US network and of a smaller network of 14 nodes were used to this aim. The obtained results lead to the conclusion that, although divide and conquer never reaches the performance level of direct optimization, it represents a valuable technique for the optimization of large networks due to the significant reduction in computation times it achieves.

1.8.3 Traffic Engineering and Resilience Based on MPLS

As already described in the previous subsection, new distributed algorithms for traffic engineering are introduced in [84, 85]. Their common characteristic is that no additional management overhead, in the form of additional intervention by network operators, is required for performing traffic engineering. However, both algorithms require enhancements to the network nodes with specific functionalities. In [87, 88, 89, 90], a different approach is preferred where traffic engineering is performed by network management based on MPLS. MPLS is a technology for establishing LSPs between node pairs in a network domain. Unlike the situation in traditional IP architectures where traffic routes are derived from the assigned link weights, the routes can be arbitrarily chosen with MPLS. This opens a large potential for traffic engineering and for novel resilience schemes, provided that a comprehensive concept is used for path establishment and network management. As no additional networking algorithms or protocols are generally needed if MPLS is used, no updates to the network technology are required. The main mechanism in most of these approaches is to use offline optimization of the LSPs. Traffic engineering methods for MPLS networks are in many respects similar to those developed for ATM networks, e.g., see [90], as both technologies essentially rely on establishing LSPs throughout the network.

Several offline traffic engineering approaches for MPLS networks are presented in [87]. In the first approach, only MPLS is used for traffic routing, and traffic engineering is performed by defining a linear program for network optimization. The network topology and traffic demand matrix provide the inputs to the linear program. The weighted sum of the average and the maximum link utilization is minimized in the objective function. For the purpose of performance evaluation, the weights are set such that either the maximum, the average, or both the maximum and average link utilization, are minimized.

Different optimization strategies are evaluated for two different networks of 20 and 25 nodes, respectively. Values of the average and maximum link utilization produced by these optimization schemes are compared to those obtained from pure IP routing. As expected, the best results in MPLS optimization are achieved when both the average and the maximum link utilization are included in the objective function. In the second approach, three different routing scenarios are compared for networks of 10, 20 and 25 nodes: unoptimized IP network, IP network with optimized link weights, and optimized MPLS network. The results show that the optimized MPLS and the optimized IP networks perform similarly well, with a slight advantage for MPLS, and that both optimized schemes by far outperform the unoptimized IP network.

As MPLS is able to simultaneously operate with IP in the same network without interference, traffic engineering possibilities with a combination of IP and MPLS traffic forwarding are also investigated. A new algorithm for traffic engineering of combined IP and MPLS networks, called Decompose-Design-Reassemble (DDR), is introduced. The input for the algorithm is a pure IP network in which a number of flows are identified as suitable for traffic engineering manipulations. Subsequently, LSPs which are supposed to accommodate these flows are computed. These LSPs will be established only when the computed path for the LSP is different from the corresponding IP path for a particular flow. This leads to a significant reduction in the number of flows actually forwarded within LSPs. The results show that, generally, only up to one third of the paths computed by such an optimization algorithm must really be routed within LSPs because of not matching the standard IP paths. This provides interesting perspectives for a significant reduction in the amount of MPLS-related state information needed in the network.

A new traffic engineering mechanism for TCP flows which uses explicit routing is introduced in [91]. The essential idea in this approach is to establish a primary and a secondary path between each pair of nodes, and to determine in real-time, on the basis of the difference in delay on the two paths, which one should be used by newly arriving flows. In analogy with RED for queue management, a set of thresholds is applied to the delay difference in order to determine the probability of using either the primary or the secondary path. Therefore, this load balancing technique is referred to as Random Early Reroute (RER). In order to avoid the "knock-on effect" known from the circuit-switched routing literature, where alternate routing is extensively used, the authors give strict priority queueing treatment to packets going through the primary paths. The proposed load balancing scheme has been implemented in the `ns-2` network simulator, and its efficiency in terms of the average per-flow goodput has been demonstrated for realistic large networks.

In contrast to traffic engineering, which is usually performed on the time-scale of minutes or hours, network capacity planning is a long-term strategic activity for network operators. Besides forecasting future traffic demands, a major factor which must be considered in the context of planning is the amount of backup capacity needed for ensuring network resilience. As the capacity installed for resilience will not be used under the normal fault-free network conditions, network operators have a strong economic incentive to minimize the amount of backup capacity installed. In [41], new protection switching mechanisms for ASs are proposed which may be implemented using MPLS. The fundamental idea is to take advantage of the load balancing potential of multipath traffic forwarding to minimize the amount of required backup capacity. Inside the AS, only disjoint border-to-border multipaths are deployed, such that network configuration and failure diagnostics are kept simple. Two basic path protection mechanisms are proposed, differing with respect to the use of the multipath structure. In the first approach, traffic is routed only on single shortest paths in the absence of failure, and multipaths are used only for backup purposes. The second mechanism first deploys multipaths in a failure-free scenario and, in case of failure, redistributes traffic from the inactive paths to the active paths. The results show that performance of the introduced protection schemes depends on the network topology, and particularly on the number of disjoint paths in the network, but not on the network size. The main metric used in performance evaluation of the presented mechanisms is the required backup capacity. It is demonstrated that the proposed schemes require significantly less backup capacity than, e.g., standard Open Shortest Path First (OSPF) rerouting. Specifically, an interesting quantitative result is that less than 20% additional resources are typically required to provide full resilience against all single link or node failures in well designed networks.

1.9 Multicast Communication

1.9.1 Protocols and Mechanisms

The popularity of large-scale distributed applications, such as video conference, multimedia dissemination, electronic stock exchange, and distributed cooperative work, has grown with the availability of high-speed networks and the expansion of the Internet. The key property of these types of applications is the need to distribute data among multiple participants while satisfying application-specific QoS requirements. This fact makes scalable multicast protocols an essential underlying communication structure. The increasing share of multicasting in high-speed data networks is expected to alter the aggregate network traffic structure and, therefore, calls for improved control mechanisms. Research is needed on the influence of QoS parameters such as reliability and scalability on network traffic.

A significant protocol element for ensuring reliability at the transport level is message loss recovery. Two representative approaches for many existing loss recovery solutions in scalable multicasting are the non-hierarchical feedback control and the hierarchical feedback control. Overall, the key issue is to reduce the number of feedback messages that are returned to the sender. In the first approach, a model that has been adopted by several wide-area applications is referred to as "feedback suppression". A well-known example is the Scalable Reliable Multicast (SRM) protocol. In SRM, when a receiver detects that it missed a message, it multicasts its feedback to the rest of the group. Multicasting feedback allows another group to suppress its own feedback. A receiver lacking a message schedules a feedback with some random delay. An improvement to enhance scalability is referred to as "local recovery", a scheme which consists in restraining message loss recovery to the region where the loss has occurred. In the second approach, hierarchical control schemes are adopted for achieving scalability for very large groups of receivers.

Scalable reliable multicast is an area of active research. In addition to the previous approaches, an alternative for ensuring reliability is Forward Error Correction (FEC). The basic idea of FEC is to predict losses and then transmit redundant data. Otherwise, recently proposed approaches based on epidemic or randomized schemes for loss recovery show promising outcomes in terms of robustness and overhead. In that direction, Bimodal Multicast (BM) provides an epidemic loss recovery mechanism: periodically, every site chooses another site at random and exchanges information to see any differences and achieve consistency. The exchanged information includes some form of message history from the group members. Epidemic protocols are simple, scale well, are robust against common failures, and eventually provide consistency as well. They combine the benefits of efficiency inherent to hierarchical data dissemination with the robustness of flooding protocols.

A large deviations approximation for evaluating the gain of using multicast over unicast in a channelized communication link is presented in [92]. This is a useful approach, especially for high-speed links, since Monte Carlo simulations have scalability problems in this case. The multicast gain can be defined as the ratio of the number of users that can be served with multicast, according to a target blocking probability, to the number of such users served with unicast. Another definition of the multicast gain, simpler to estimate, is the ratio of the number of users that generate the mean link occupancy in each communication method (multicast and unicast). The idea in the combination of unicast and broadcast consists in using broadcast to transmit the most popular content. In order to make this work, the operator needs to select the appropriate content and to allocate the optimal number of channels that minimizes the blocking probability of unicast. The gain over unicast is similarly computed with a combination of unicast and broadcast instead of multicast. In this work, the individual user channel preference is modeled by a Zipf law, assessed in the literature to be a good representation of human behavior in this respect. The study shows that, for large values of the Zipf distribution parameter, the combination of unicast and broadcast outperforms the use of multicast. The results are based on approximations and rely on the assumption that the operator is able to select the most popular channels. Thus, on the whole, broadcasting may be considered as a viable alternative, all the more because of the practical problems of implementing the multicast functionality in the networks.

1.9.2 Performance and Traffic Properties

Both contributions [93] and [94] focus on the traffic properties of BM, that are induced by its epidemic recovery mechanism, in comparison to SRM. The empirical results demonstrate that BM generates more desirable traffic than SRM. The studies elaborate on the protocol mechanisms as the underlying factor explaining these results. From a traffic perspective (further details about it can be found in Section 3.3.1), these studies aim at discovering and developing multicast protocols that not only feed well-behaved traffic discreetly into the networks, but also can cope with the existing self-similar traffic and its adverse consequences.

[95] considers the epidemic anti-entropy model of BM, used to ensure protocol reliability, and demonstrates its scalability and robustness. Comparative simulation results discussing the model performance on a range of typical scenarios are given. It is demonstrated that anti-entropy loss recovery produces balanced overhead distribution among group receivers and is scalable with respect to an increase in group size, multicast message rate, or system-wide noise rate. The anti-entropy distribution model provides eventual data consistency, scalability, and independence from communication topologies. It also handles network failures quite transparently and is able to work with minimal connectivity.

As a continuation of above studies, [96] focuses on the use of epidemic techniques in the context of a reliable and scalable multicast protocol for ad-hoc networks. Being dedicated to ad-hoc networks, this contribution is mainly described in Section 6.4.3, where further developments can be found. Let us just state here that the non-deterministic nature and the peer-based communication style in epidemic algorithms are well matched to the requirements of ad-hoc networks. The study includes a brief survey of epidemic algorithms for reliable multicasting in ad-hoc networks, some analytical results for simple epidemics, and the development of a Peer-to-Peer (P2P) anti-entropy algorithm for content distribution together with a prototype simulation model.

Overall, the results of these activities on multicast can be viewed to fit in the general problem of integrating multicast communication into the Internet. In particular, mechanisms including efficient loss recovery and scalability to support multicast transport would be essential. Various applications on both wired and wireless ad-hoc settings could benefit from multicast communication support with efficient provision of reliability guarantees.

1.10 Graph Models for Interconnection Networks

Some recent works have been motivated by the search for an adequate characterization of the topology of huge networks like the global Internet, considered at resolutions ranging from router level to AS level, or various virtual networks built on it, like P2P networks. A necessary prerequisite for those studies is provided by empirical results on Internet topology obtained from systematic large-scale sensing with tools such as `traceroute`. Organizations like the University of Oregon, with its Route Views Project, and the Cooperative Association for Internet Data Analysis (CAIDA) have played a central role in this activity. The Web sites of these organizations can be found at `http://www.routeviews.org/` and `http://www.caida.org/`.

One of the most interesting models studied in this context is the so called Power-Law Random Graph (PLRG). In this model, the node degrees are independently drawn from a power-law distribution, following which the connections are randomly chosen. This model is extremely simple, so it is somewhat surprising that its resemblance to the real Internet is much better than one might have thought. Four COST 279 contributions have been devoted to this model or a modification of it; it should be noted that these papers focus on the model itself, not on the current Internet topology.

The main result in papers [97, 98] is the following. First, denote by N the number of nodes and assume the tail of the degree distribution decreases as $P(D > d) \sim d^{-\tau+1}$, where $\tau \in (2, 3)$, thereby possessing infinite variance. Then, the hop distance between two randomly selected nodes of the giant component (the random graph does not necessarily need to be fully connected) is asymptotically almost surely less than $(2 + o(1))k^*$, where

$$k^* = \left\lceil \frac{\log \log N}{-\log(\tau - 2)} \right\rceil .$$

Thus, the practical diameter of the graph hardly grows at all when N runs through many orders of magnitude. This phenomenon, and the proof of the theorem, is based on a kind of "soft hierarchy" where nodes with high degrees form a "core network" providing short connections. The degree threshold above which a node is counted as belonging to this core network is an increasing function of N asymptotically smaller than any positive power of N. This result has been recently improved, outside COST 279, in [99, 100, 101], where asymptotic distributions of the distance were derived in all three main parameter regions $\tau \in (1, 2)$, $\tau \in (2, 3)$, and $\tau > 3$. Three qualitatively different asymptotic connectivity structures, or "architectures", are obtained. On one hand, with $\tau \in (1, 2)$, the core network is fully connected and all nodes of the giant component have a direct link to it. On the other hand, the case $\tau > 3$ provides an asymptotically "flat" architecture where the distances are typically of order $\log N$.

The main contribution in [102] is the "conditionally Poissonian random graph", a modification of the PLRG model that has even nicer mathematical properties. Instead of independently drawing the degrees D_i from a chosen power-tailed distribution, only independent identically distributed (i.i.d.) "capacities" Λ_i of the nodes are drawn first. Then, each pair of nodes (i, j) is connected, independently from other pairs, with a number of links that follows a Poisson distribution with mean

$$\frac{\Lambda_i \Lambda_j}{\sum_{k=1}^{N} \Lambda_k}.$$

This definition is in fact a slight modification of the "expected degree sequence" model in [103]. Thanks to the summation rule of independent Poisson random variables, the distribution of the number of links joining any two disjoint sets of nodes is Poisson again. The proofs in [98] can then be carried out in a more elegant way.

The paper [104] performs further studies on the PLRG model, mainly based upon simulations. A routing system is added to the core part of the model by assuming some "natural" routing principle, where the routing is hereditary, invertible, and makes use of least-hop paths. This lets the traffic concentrate very unevenly on the different links. Starting with a uniform traffic matrix, a roughly exponential distribution of link loads can be observed in this set-up.

Since a small minority of large nodes, i.e., with a high degree, plays a central role in the high connectivity level of the PLRG model, it was interesting to study the vulnerability of PLRG regarding possible outages of large nodes. When only the huge nodes with degree higher than $\sqrt{N}$ were removed, the node distances became a little bit longer, but the effect was not dramatic. When the whole core was lost, a large part of the network still remained connected but, now, the distances became substantially longer. These results tell of a very high resilience level of the topology, but they are probably over-optimistic in comparison with the real Internet.

The last topic studied in [104] was the implications of the model to the efficiency of multicasting. It was found that in a moderate size multicast tree between randomly selected nodes, most of the gain is obtained in the portion of the network between the source and the center part of the core, after which the routes towards different receivers do not overlap much. This rather clearly differs from the more standard scenario of branching resembling a binary tree.

Chapter 2
Queueing Models

Sabine Wittevrongel
Ghent University, Belgium

Contributors:

Samuli Aalto (Helsinki University of Technology, Finland), Nail Akar (Bilkent University, Turkey), Carlos Belo (Telecommunications Institute, Portugal), Hans van den Berg (TNO Telecom, Netherlands), Markus Fiedler (Blekinge Institute of Technology, Sweden), Dieter Fiems (Ghent University, Belgium), Peixia Gao (Ghent University, Belgium), Veronique Inghelbrecht (Ghent University, Belgium), Robert Janowski (Warsaw University of Technology, Poland), Udo Krieger (Otto Friedrich University Bamberg, Germany), Koenraad Laevens (Ghent University, Belgium), Tom Maertens (Ghent University, Belgium), Michel Mandjes (Center for Mathematics and Computer Science, Netherlands), Ilkka Norros (VTT Information Technology, Finland), Olav Østerbø (Telenor Research & Development, Norway), Detlef Sass (University of Stuttgart, Germany), Ana da Silva Soares (Université Libre de Bruxelles, Belgium), Kathleen Spaey (University of Antwerp, Belgium), Hung Tran (Telecommunications Research Center Vienna, Austria), Jorma Virtamo (Helsinki University of Technology, Finland), Stijn De Vuyst (Ghent University, Belgium), Joris Walraevens (Ghent University, Belgium), Sabine Wittevrongel (Ghent University, Belgium)

2.1 Introduction

This chapter presents an overview of queueing models studied within COST Action 279. Such models are important tools to investigate the behavior of the buffers used in various subsystems of telecommunication networks, and hence to evaluate the quality of service, in terms of loss and delay, experienced by the users of a communication network. In Section 2.2 a number of discrete-time queueing models are discussed. Section 2.3 addresses some new developments with respect to fluid flow analysis. Work on Gaussian storages is reported in Section 2.4. In Section 2.5 some new results on processor sharing models are

J. Brazio et al. (eds.), Analysis and Design of Advanced Multiservice Networks Supporting Mobility, Multimedia, and Internetworking, 55–84.

presented. Section 2.6 discusses recent work on multilevel processor sharing models. Section 2.7 is devoted to the analysis of a variety of other continuous-time queueing models. Some techniques to study end-to-end delays in networks of queues are described in Section 2.8. Finally, Section 2.9 overviews some specific models and analysis techniques for performance evaluation in the context of optical networks.

2.2 Discrete-Time Queueing Models

In a discrete-time queueing model the time axis is assumed to be divided into fixed-length intervals, usually referred to as *slots*. This section provides an overview of a number of specific discrete-time queueing models studied within COST 279, as well as results obtained for these models. For the analysis of the models, analytical techniques mainly based on an extensive use of Probability Generating Functions (PGF) have been developed.

2.2.1 Queues with Static Priority Scheduling

Priority scheduling is a hot topic in multimedia networks. For real-time applications, it is important that the mean packet delay and delay jitter are not too large. Therefore, this type of traffic is given priority over non-real-time traffic in the switches/routers of the network, i.e., delay-insensitive traffic is serviced in a switch only when no delay-sensitive traffic is present.

In [105] is considered a discrete-time single-server queueing system with infinite buffer space, a Head-of-Line (HOL) priority scheduling discipline, and a general number of priority classes. All types of packet arrivals are assumed to be independent and identically distributed (iid) from slot to slot, but within one slot the numbers of packet arrivals from different classes can be correlated. The system has one server that provides the transmission of packets at a rate of one packet per slot, i.e., the service times of the packets are deterministically equal to one slot. First, an expression for the joint PGF of the system contents of all priority classes is derived. From this joint PGF, the marginal PGFs of the system contents of each priority class separately and of the total system content are found. Furthermore, the PGFs of the packet delays of each class are calculated. The analysis of the latter is largely based on the concept of *sub-busy periods*. From the generating functions obtained, performance measures, such as the moments and approximate tail probabilities of system content and packet delay, are derived. Especially the analysis of the tail behavior is an important result of [105]. It is shown that the tail behavior is not necessarily exponential. Figure 2.1 illustrates the effect of HOL priority scheduling in a 16×16 output

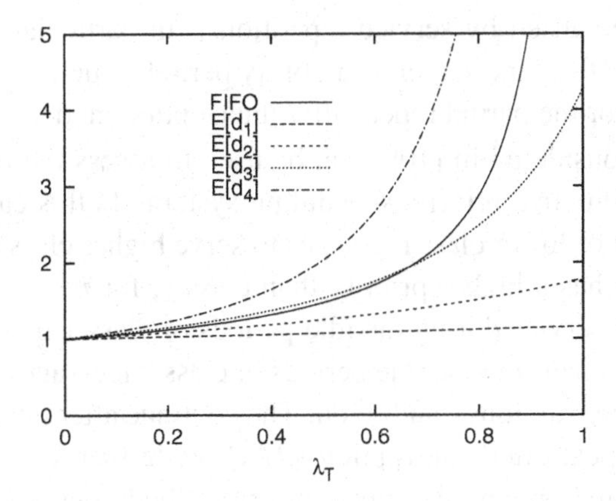

Figure 2.1: Mean delay for each priority class versus the total arrival rate λ_T in case of HOL priority and FIFO

queueing switch with independent and uniform routing. Packet arrivals on each inlet are independent and generated by a Bernoulli process. There are 4 priority classes and each class corresponds to a fraction of 0.25 in the overall traffic mix. The figure shows the mean packet delay in an output queue for each priority class versus the total arrival rate, together with the mean packet delay in the case of First-In-First-Out (FIFO) scheduling. For two priority classes, the analysis of [105] has also been extended to the case of general packet service times. Specifically, in [106, 107], general packet service times and a preemptive priority discipline are considered, whereas [108] considers general service times and non-preemptive priorities.

Another way to study queues with priority scheduling is discussed in [109]. In this paper, a discrete-time single-server queue subjected to *server interruptions* is investigated. Server interruptions are modeled as an on/off process with geometrically distributed on-periods and generally distributed off-periods. The messages that arrive in the system possibly require more than one slot of service time, implying that a service interruption can occur while a message is in service. Therefore, different operation modes are considered, depending on whether service of an interrupted message continues, partially restarts, or completely restarts after an interruption. For each alternative, expressions for the steady-state PGFs of the buffer contents at message departure times and at random slot boundaries, of the unfinished work at random slot boundaries, of the message delay, and of the lengths of the idle and busy periods are established. From these results, closed-form expressions for various performance measures, such as means and variances of the buffer occupancy and of the message delay, are derived. Numerical results show the deterioration of system performance caused by service repetitions. In particular, it is observed that the mean length of the server availability periods crucially determines the system stability for the partial repeat after interruption mode.

The model considered in [109] can be used to assess the performance of a multi-class preemptive priority scheduling system. In this case, the system interrupts service of lower class messages to serve higher class messages. Assume that class i has a higher priority than class j, for $i < j$. Then, class 2 messages receive service during the idle periods of class 1 messages. Class 3 messages are served during the idle periods of class 2 messages (the busy periods include the interruptions), and so on. The continue after interruption (CAI) mode and the repeat after interruption (RAI) mode then correspond to preemptive resume and preemptive repeat priority scheduling, respectively. The partial repeat after interruption (p-RAI) operation mode can be considered as an intermediate case between both types of priority scheduling and allows the study of the influence of packetizing in preemptive priority systems.

2.2.2 Queues with Dynamic Priority Scheduling

In a static priority scheme, as discussed above, priority is *always* given to the delay-sensitive class, and thus packets of this class are always scheduled for service before delay-insensitive packets. It has been shown that this scheme provides relatively low delays for the delay-sensitive class. However, if a large portion of the network traffic consists of high-priority traffic, the performance for the low-priority traffic can be severely degraded. Specifically, HOL priority scheduling can cause excessive delays for the low-priority class, especially if the network is highly loaded. This drawback is also known as the *starvation problem*. In order to find a solution for this problem, several dynamic priority schemes have been proposed in the literature. These schemes are mostly obtained by alternately serving high-priority traffic and low-priority traffic, depending on a certain threshold, or by allowing priority jumps. In the latter type, referred to as head-of-line with priority jumps (HOL-PJ), when high- and low-priority packets arrive at the respective queues, packets of the low-priority queue can jump to the high-priority one, as illustrated in Figure 2.2. Many criteria can be used to decide if and when low-priority packets jump: a maximum queueing delay in the low-priority queue, a queue-length threshold of the low-priority queue, or a random jumping probability per slot. Further, the jumping process is also characterized by the number of packets that jump at the same time and by the specific moments when these packets jump, e.g., at the beginning of a slot, or at the end of a time slot.

In [110] and [111], a discrete-time single-server two-class queueing system

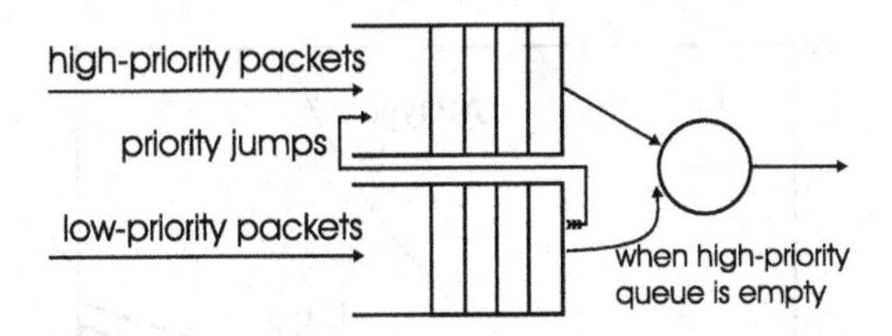

Figure 2.2: Two-class single-server queue with HOL-PJ priority scheduling

with infinite buffer size and HOL-PJ priority scheduling is considered. Two types of packets thus arrive in the system, and these two classes are assumed to arrive in separate, logical queues, i.e., a high- and a low-priority queue. The numbers of arrivals of both classes are assumed to be iid from slot to slot. However, within one slot, the numbers of arrivals from both classes can be correlated. The service times of the packets are equal to one slot. Whenever there are packets in the high-priority queue, they have service priority, and only when this queue is empty can packets in the low-priority queue be served. Within a queue, the service discipline is FIFO.

The difference between the models studied in [110] and [111] lies in the jumping process. In [110], the total content of the low-priority queue jumps with a constant probability β to the high-priority queue in each slot, while in [111], only the packet at the HOL-position of the low-priority queue can jump to the high-priority queue. The possible jump in [111] depends on the content of the high-priority queue, i.e., only when that queue is not empty, the packet jumps with probability one. When the high-priority queue is empty, the low-priority packet at the HOL-position is immediately served. For both models, an expression for the joint PGF of the system contents of the high- and low-priority queues is obtained. From this joint PGF, the marginal PGFs of the system contents of each queue separately and of the total system content are derived. Also, the PGFs of the packet delays of both classes are calculated. From the obtained PGFs, performance measures, such as mean values and variances, are found. Numerical results show the impact and significance of both investigated dynamic priority scheduling disciplines in an output queueing switch.

2.2.3 Queues with Place Reservation

In [112], a new kind of priority discipline is studied that provides a better Quality of Service (QoS) to packets that are delay-sensitive at the cost of allowing higher delays for best-effort packets. The idea, first suggested in [113], is to

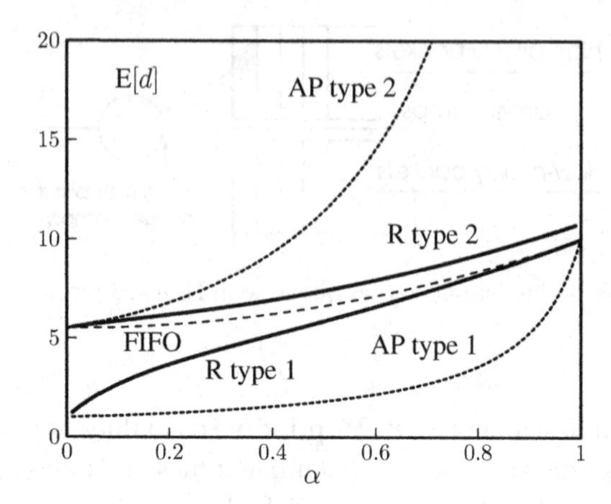

Figure 2.3: Mean delay for both types of packets versus the fraction α of type 1 traffic in case of place reservation (R), absolute priority (AP) and FIFO

introduce a reserved space in the queue that acts as a place holder for future arriving high-priority packets. A discrete-time queue is considered with arrivals of type 1 (delay-sensitive) and type 2 (best-effort). Whenever a packet of type 1 enters the queue, it takes the position of the reservation that was created by a previous arrival of that type and creates a new reservation at the end of the queue. On the other hand, type 2 arrivals always take place at the end of the queue in the usual way. In this way, a packet of type 1 may jump over one or more packets of type 2 when being stored in the queue, thus lowering its delay compared to packets of type 2. In [112], the exact PGFs of the delays experienced by both types of packets are obtained and, from these PGFs, mean values, variances and tail probabilities are calculated.

Figure 2.3 illustrates the delay differentiation between the two types of packets under the considered reservation discipline. The numbers of arrivals per slot of type 1 and type 2 are independent and have a geometric and a Poisson distribution, respectively. The total arrival rate is $\lambda_T = 0.9$, and the mean delays for packets of type 1 and type 2 are plotted versus the fraction α of type 1 packets in the overall traffic mix. The values obtained for FIFO and both types of packets under absolute priority (AP), also known as HOL priority, are shown as well. Note that, in the case of FIFO, it is considered the delay of an arbitrary packet *regardless* of its type. It is observed that under the reservation discipline, type 2 packets are less likely to experience an extremely large delay than in the case of absolute priority.

2.2.4 Multiserver Queues

In most of the existing literature on discrete-time multiserver queueing models, the service times of customers are assumed to be constant, equal to one slot or multiple slots. In [114], a discrete-time multiserver queue with geometric service times, an infinite buffer size, a First-Come-First-Served (FCFS) service discipline, and general independent packet arrivals is considered. The behavior of the queueing system is studied analytically by means of a generating-functions approach. This results in closed-form expressions for the PGFs of the system content and the packet delay. From these PGFs, expressions are obtained for performance measures such as the mean values, variances, and tail probabilities of the system content and the delay. In [115], the analysis is further extended from the case of an uncorrelated packet arrival process to the case of a two-state correlated arrival process. The delay analysis is based on a general relationship between system content and packet delay established in [116], valid for any discrete-time multiserver system with geometric service times, regardless of the exact nature of the arrival process.

In [117], a discrete-time multiserver queueing system with preemptive resume priority scheduling is investigated. Two classis of traffic are considered; the first class has preemptive resume priority over the second one. The service times are again assumed to be geometrically distributed, but now with a rate dependent on the traffic type. An expression for the steady-state joint PGF of the system contents of the high- and the low-priority traffic is derived. From this, closed-form expressions for the PGFs and the moments of the system contents, both for the high and low priority traffic, can be obtained. By means of Little's law, the mean delay for the two types of traffic can then also be found.

2.2.5 Queues with Server Vacations

Queueing systems with server vacations have proven to be a useful abstraction of systems where several classes of customers share a common resource, such as priority systems and polling systems, or of systems where this resource is unreliable, such as maintenance models and Automatic Repeat Request (ARQ) systems. A discrete-time gated vacation system is considered in [118]. The classical gated vacation system can be seen as one with two queues separated by a gate. Arrivals are routed to the queue before the gate, whereas customers in the queue after the gate are served on a FIFO basis. When there are no more customers in the latter queue, the server takes a vacation and opens the gate—all customers move instantaneously from the queue before the gate to the queue after it—upon returning from this vacation. In [118], the classical gated vacation queueing system is extended in the sense that customer arrivals are also

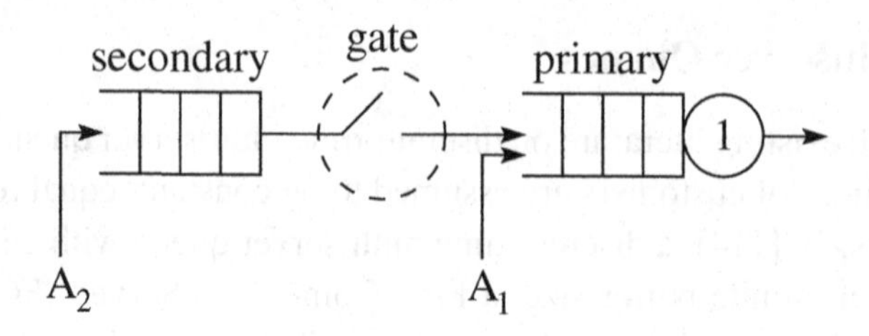

Figure 2.4: Gated-exhaustive vacation system

allowed to be immediately routed to the queue after the gate, as illustrated in Figure 2.4. The model under investigation allows to capture the performance of, among others, the exhaustive (only arrivals in primary queue) and the gated (only arrivals in secondary queue) queueing systems with multiple or single vacations. Using a generating-functions approach and the method of the supplementary variable, expressions are obtained for performance measures such as the moments of the system content at various epochs in equilibrium and of the customer delay. The results depend on a constant that has to be determined numerically.

2.2.6 Queues in ARQ Systems

ARQ protocols are used to obtain reliable transfer of packets from a sender to a receiver communicating over an unreliable channel, where packet corruption or loss is possible. In [119], an analytical approach for studying the queue length and the packet delay in the transmitter buffer of a system working under a stop-and-wait retransmission protocol is presented. The operation of the stop-and-wait ARQ protocol is illustrated in Figure 2.5. The transmitter sends a packet available in its queue and then waits for s slots (the feedback delay) until it receives the corresponding feedback message. If the packet was transmitted correctly (positive acknowledgement or ACK), the next packet waiting in the queue is transmitted. Otherwise, if an error occurred (negative acknowledgement or NACK), the packet is retransmitted. The buffer at the transmitter side is modeled as a discrete-time queue with an infinite storage capacity. The numbers of packets entering the buffer during consecutive slots are assumed to be iid random variables. The information packets are sent through an unreliable and non-stationary channel, modeled by means of a two-state Markov chain. An explicit formula is derived for the PGF of the system content. This PGF is then used to derive several queue-length characteristics as well as the mean packet delay.

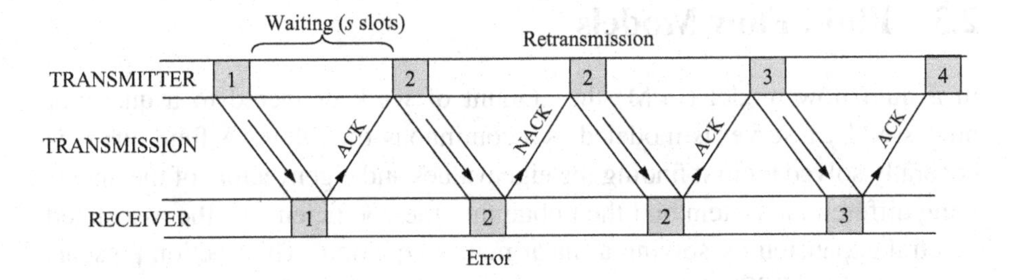

Figure 2.5: Stop-and-wait ARQ protocol

The model was studied further in [120], where not only the mean value but the whole distribution of the packet delay is obtained, as well as an expression for the maximum throughput of the system. For both, the analysis is based on the *conditional effective service times*, i.e., the time that elapses from a packet's first transmission until the transmitter is notified of the correct receipt of that packet.

2.2.7 Queues with Bursty Traffic

Two discrete-time queueing models are compared in [121]: a *packet model*, where two timescales, a burst and a packet timescale, are present in the input traffic, and a *fluid model*, where only an (identical) burst timescale is present. The time axis is assumed to be divided in fixed-length time units, called frame times, and every frame time is further divided into fixed-length units, called packet times. A two-state discrete-time Markov source is considered, which has a frame time as underlying time unit and thus can only change state at frame time boundaries. During a frame time in which the source is in the first, respectively second state, it generates a certain amount of bytes. In the packet model, these bytes are divided into fixed-length packets that are sent all together in the beginning of the frame time. In the fluid model the bytes are sent at a constant rate over the whole duration of the frame time. A queueing model with an infinite buffer capacity is then considered in which the traffic of either M identical packet sources, or M identical fluid sources with the same parameters, is multiplexed. For both systems, the distribution of the amount of unfinished work is derived, and the impact of approximating packetized flows by fluid flows on the complementary cumulative distribution (CCD) of the unfinished work is investigated. The main conclusions are described in Section 3.4.5.

2.3 Fluid Flow Models

In a fluid flow model (FFM) the amount of work delivered to a queue or processed by a server is modeled as a continuous-time flow. A fluid queue is generally solved by first finding the eigenvalues and eigenvectors of the underlying differential system and then obtaining the coefficients of the associated spectral expansion by solving a linear matrix equation. This section presents some new COST 279 developments with respect to fluid flow analysis.

2.3.1 Algorithmic Approach

Consider a Markov modulated fluid queue, i.e., a two-dimensional continuous-time Markov process $\{(X(t), \varphi(t)) : t \in \mathbb{R}^+\}$ where $X(t)$ takes values in $\mathbb{R}^+$ and $\varphi(t)$ in $\mathcal{S}$, a finite set. The component $X(\cdot)$ is called the *level* and $\varphi(\cdot)$ is called the *phase*. The level is subordinated to the phase in the following way. The phase process $\{\varphi(t) : t \in \mathbb{R}^+\}$ is an irreducible Markovian process. During those intervals of time in which the phase is constant, say equal to i, the level increases or decreases at a constant rate dependent on i, or it remains constant. If $X(t) = 0$ and if the rate at time t is negative, then the level remains at 0.

Ramaswami [122] shows that $\{X(t)\}$ has a phase-type stationary distribution using the dual process of $\{(X(t), \varphi(t))\}$. Most importantly, he also constructs a very efficient computational procedure, based on the logarithmic-reduction algorithm of Latouche and Ramaswami [123] for discrete-level Quasi-Birth-Death (QBD) processes: he thereby reduces a complex continuous time, continuous state space problem to a familiar, simple discrete time, discrete state space system. The use of the dual process in [122] is motivated by a property that relates the stationary distribution of the original $\{(X(t), \varphi(t))\}$ process to first passage probabilities at level 0 in the dual process. In [124], the similarities with QBDs are reinforced by showing that one may actually *directly* use these first passage probabilities in the original process. Also, another probabilistic interpretation of Ramaswami's computational procedure is given.

2.3.2 Large-Scale Finite Fluid Queues

Except for some structured models, e.g., the Anick-Mitra-Sondhi (AMS) fluid flow model [125], it is in general hard to find the eigensystem in a computationally efficient and stable manner. Moreover, the linear matrix equation to solve for the coefficients in the finite fluid queue case is known to be ill-conditioned, especially in the case of large buffer sizes.

In [126], a numerically efficient and stable method for solving large-scale finite Markov fluid queues is developed. No special structure is imposed on the underlying continuous-time Markov chain, i.e., the eigenvalues and eigenvectors need not be determined in closed form. The authors propose an alternative method that relies on decomposing the differential system into its stable (forward) and anti-stable (backward) subsystems, as opposed to finding eigenvalues, using a method based on the additive decomposition of a matrix pair with respect to the imaginary axis. There are a variety of numerical linear algebra techniques, with publicly available codes, that can be used for such an additive decomposition, including the generalized Newton iterations, the generalized Schur decomposition, and the spectral divide and conquer methods. Using the generalized Newton iterations, which have quadratic convergence rates, it is shown in [126] that the accuracy of the proposed method does not depend on the buffer sizes and that in the limit the finite fluid queue solution converges to that of the infinite fluid queue. Moreover, it is demonstrated that fluid queues with thousands of states are efficiently solvable using the proposed algorithm.

2.3.3 Voice and Multi-Fractal Data Traffic

The FFM has shown being able to incorporate many types of traffic, i.e., superpose them for analysis in a unified model. In [127], it is augmented by a model displaying multi-fractal behavior, described in Section 3.4.6. This model can be matched to the characteristics of real traffic by choosing the appropriate parameters for the sub-processes. The paper investigates the interaction between multi-fractal data traffic and voice traffic with suppressed silence phases and consideration of the on-hook-state of the Internet phone, the model for the latter being a 3-state ON-OFF model. The fluid flow calculations can be used in a straightforward manner, with the only exception that the pseudo-rates of the sub-processes contributing to the data traffic process are multiplied instead of added. As a result, formulas expressing queuing delay and loss as experienced individually by voice and data are obtained. A case study is carried out, investigating the maximal load under given delay quantiles. As expected, this load level depends heavily on the variance of the data traffic. In general, the voice traffic yields better performance in terms of loss and delay and helps to increase the maximal load, while still meeting the target performance values.

2.3.4 Superposition of General ON-OFF Sources

The effect of a superposition of general ON-OFF sources on a multiplexer is studied in [128]. Sources are statistically identical and independent. During the ON period a source emits at a constant rate, either in a fluid flow fashion, or by periodically emitting fixed size packets. During the OFF period the source remains silent. Both the ON and OFF periods are random variables with general distributions but finite mean values. The distributions considered include the Pareto type, known to lead to traffic having the long range dependence (LRD) property.

The superposition of a number of such general ON-OFF sources results in a stochastic process called semi-birth-and-death (semi-BD). The state of the semi-BD process is the number of active sources at a given time; the random variable of interest is the holding time in that state. In this case, the traffic generated simply equals the number of active sources multiplied by the individual rate. For the case of the semi-BD, the stationary distribution of the holding time is given in [128] in explicit form as a function of the state considered, the number of sources, and the distribution of the ON and OFF periods. It is further argued that, for the semi-BD, the distribution of the holding time would tend to an exponential even with a moderate number of sources. The above result strongly suggests that the now classical AMS solution for the probability of buffer overflow in an infinite buffer multiplexer with a superposition of exponential ON-OFF sources as input could be applied to the case of general ON-OFF distributions. The remaining of [128] is devoted to evaluating, theoretically and by way of simulation, the circumstances under which the AMS solution is a good approximation as a function of the number of sources, the distributions of the ON and OFF periods, and the desired level of probability of overflow.

2.3.5 Feedback Fluid Queues

In [129] is considered a single point in an access network where several bursty users are multiplexed. The users adapt their sending rates based on feedback from the access multiplexer. Important parameters are the user's peak transmission rate p, which is the access line speed, the user's guaranteed minimum rate r, and the bound ϵ on the fraction of lost data. Two feedback schemes are proposed and studied. In both schemes the users are allowed to send at rate p if the system is relatively lightly loaded, at rate r during periods of congestion, and at a rate between r and p, in an intermediate region. For both feedback schemes an exact analysis is presented, under the assumption that the users' file sizes and think times have exponential distributions. The techniques are used to design the schemes jointly with admission control, i.e., the selection of the number of admissible users, to maximize throughput for given p, r, and ϵ. Next is considered the case where the number of users is large. Under a specific (many-sources) scaling, explicit large deviations asymptotics for both models are derived. The extension to general distributions of user data and think times is discussed.

The model is also extended to a "buffer-dependent" Markov fluid queue, defined as follows. A Markov fluid source is defined as a continuous-time Markov chain with transition rate matrix Q of dimension d, and a traffic rate vector r, describing the generation of traffic at a constant fluid rate $r(i)$ when the Markov chain is in state i, for $i = 1, \ldots, d$. Now consider N of these Markov fluid sources of dimension d, characterized by the pairs (Q_n, r_n), for $n = 1, \ldots, N$, and suppose traffic is generated according to the nth Markov fluid source when the buffer level is between B_{n-1} and B_n, where the B_n are increasing, with $B_0 = 0$ and B_N finite or infinite. For this model, the complete buffer content distribution is derived in terms of the solution of an eigensystem.

2.3.6 Fair Queueing Systems

In [130], an FFM for a fair queueing system with unidirectional coupling for several different classes is considered, where each class has a predefined minimum bandwidth guaranteed. These minimum bandwidths form a decomposition of the total bandwidth and avoid starvation of a class caused by another class. A multiplexing gain is achieved by passing down the residual bandwidth of a class to the lower adjacent class. Therefore, this system is called *unidirectionally coupled.* The case of a unidirectionally coupled fair queueing system consisting of two classes is shown in Figure 2.6. Also, each class has its own FIFO buffer for exclusive usage. The system is described by an FFM, and hence sources and server are assumed to be Markov modulated fluid processes (MMFP). The key observation concerning the residual bandwidth is that a class lending its residual bandwidth is oblivious to this. Also receiving residual bandwidth is, from the receiving class point of view, just an additional server process. This process is interpreted to be, again, an MMFP, and certain states and the buffer's mean busy period of the lending adjacent class are used to model this additional process. The states mentioned are the underload states, because state transition among these states reflects—assuming the buffer to be empty—the dynamics of the residual bandwidth. The transition matrix of the additional server process can be built with it. The distribution of the buffer content is found by applying the FFM. To determine the distribution for a specific buffer, the distribution of the previous adjacent buffer has to be calculated, apart from that of the first buffer. Also, two estimates for the overflow probability of a buffer are obtained. First, a more accurate estimate is calculated by applying the full FFM, i.e., all eigenvalues and eigenvectors are used. Second, a fast and robust estimate is found by using only the domi-

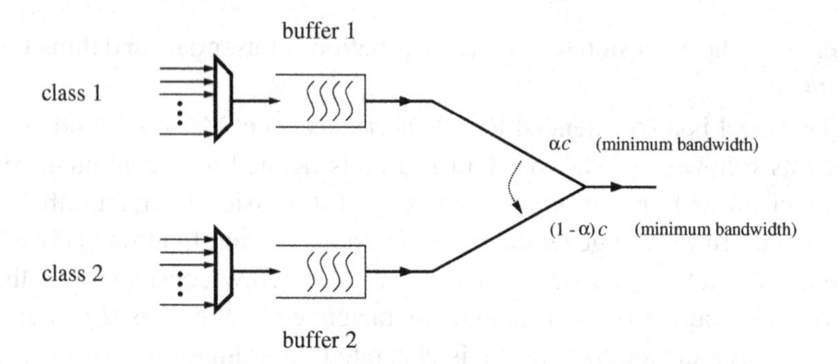

Figure 2.6: Example of a unidirectionally coupled fair queueing system

nant eigenvalue and the Chernov large deviation bound. This is a conservative estimate with larger deviation than the other estimate.

2.3.7 Bottleneck Identification and Classification

The stochastic FFM [125] has also shown to be capable of revealing the impact of bottlenecks, i.e., shortages in capacity, on packet streams by comparing bit rate histograms at the output with those at the input of the bottleneck [131]. From standard fluid flow analysis for Markov Modulated Rate Processes (MMRP), the joint probabilities that the buffer is empty, or non-empty, in each state are calculated. As shown in [131], these probabilities are the key for obtaining the output bit rate distribution both for individual and total traffic streams. From comparisons with input bit rate distributions, it can be seen whether there is interfering traffic in the bottleneck or whether the bottleneck has a buffer of significant size. This allows not only for an identification, but also for a classification of the bottleneck. While the maximal capacity of the bottleneck is revealed in the output bit rate histogram of the total stream, it may under certain conditions also become visible in the corresponding histogram of an individual stream.

2.4 Gaussian Storages

This section overviews some new developments with respect to the most probable path technique to derive estimates of the queueing performance for queues with Gaussian input traffic. Also, a method to determine delay quantiles of a multiplexer with Gaussian input, involving a fitting procedure to Ornstein-Uhlenbeck processes, is discussed.

2.4.1 Most Probable Path Technique

By the theory of large deviations of Gaussian processes, the probability that a simple queue with Gaussian input exceeds a level x can be approximated by

$$\Pr[Q \geq x] \approx \exp\left(-\frac{1}{2}\|f_x\|_R^2\right), \tag{2.1}$$

where f_x is the path of the input process that creates a queue of size x at time 0 and has the smallest reproducing kernel Hilbert space (RKHS) norm $\|\cdot\|_R^2$ among all paths creating such a queue.

The framework was generalized in a straightforward way to a two-class priority system in [132] as follows. Assume that the two arrival processes are independent continuous Gaussian processes with stationary increments. Consider the most probable *path pair* that creates a total queue (both classes together) of size x. If this path of the higher priority input does not create a positive queue at time 0, then this path pair is also the most probable one to create a lower class queue of size x.

The remaining case is studied in [133]. The idea is to compute an easily characterized heuristic approximation to the most probable path pair, where the higher priority traffic essentially fills the link while the lower class traffic is accumulating in the queue. The same principles can be applied to a generalized processor sharing (GPS) system with two classes by replacing link filling by filling the quota guaranteed for a traffic class. Simulations show that these approximations are sufficiently accurate for many practical purposes, such as studying the effects of setting GPS weights.

In [134], another kind of modification of the basic Gaussian queue is studied. The service capacity is not any more constant, but continuously varied according to the traffic rate observed, with a constant delay Δ. The allocated capacity is $1+\epsilon$ times the observed rate. That is, the cumulative service process is defined as

$$C_t = (1+\epsilon)(A_{t-\Delta} - A_{-\Delta}), \tag{2.2}$$

where A is the cumulative input process. The queue length process is defined as a supremum of the net input process:

$$Q_t = \sup_{s \leq t}((A_t - A_s) - (C_t - C_s)). \tag{2.3}$$

Since the net input process is Gaussian, the basic estimates of the queue length distribution and of the most probable paths are directly available.

Figure 2.7 shows, in case of fractional Brownian motion (fBm) input, the most probable path that creates a queue of size 4 at time 0. Note how the input

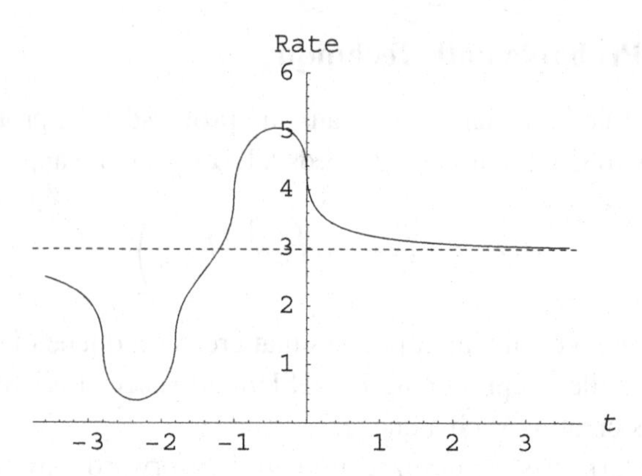

Figure 2.7: The input rate of the most probable way to obtain a queue of size 4, when service capacity is varying according to traffic prediction. The input process is a fBm with Hurst parameter 0.75

process "fools" the prediction by making the input first very slow and then, when the control cannot react any more, suddenly speeding up.

In [135] is made a substantial contribution to the most probable path approach described above by the establishment of a technique for identifying most probable paths that are truly infinite-dimensional combinations of covariance functions. This paper focuses on the simplest of this kind of problems. Consider a simple Gaussian queue with centered input process Z and unit service rate. What path β^* is the most probable one, in the sense of smallest RKHS norm, among input paths f that produce a busy period starting at 0 and straddling the interval $[0, 1]$, that is, satisfying $f(t) \geq t$ for all $t \in [0, 1]$? This event is an intersection of (infinitely many!) half-spaces and thus convex. Assume that it is also non-empty, very mild conditions being sufficient for this. Then, β^* exists and is unique. Denote $S^* = \{s \in [0, 1] \,|\, \beta^*(s) = s\}$. The crucial observation is that

$$\beta^* \in \bigcap_{\epsilon > 0} \overline{\text{sp}}\{Z_u \,|\, u \in [0, 1],\ d(u, S^*) < \epsilon\}, \tag{2.4}$$

where $d(u, S^*) = \inf\{|u - s| \,|\, s \in S^*\}$ and $\overline{\text{sp}}$ denotes the closure of linear span. This result makes it often possible to identify a most probable path numerically and in some cases even analytically. When the process Z is non-smooth, for example fBm, we have $\beta^* \in \overline{\text{sp}}\{Z_u \,|\, u \in S^*\}$, and $\beta^*(t) = \mathbb{E}[Z_t \,|\, Z_s = s\ \forall s \in S^*]$. For Brownian motion, it is well known that $S^* = \{1\}$. In the case of fBm, it turns out that S^* has the form $[0, s^*] \cup \{1\}$ (resp. $[s^*, 1]$) if the self-similarity parameter H is larger (resp. smaller) than $1/2$.

When Z is smooth, a characterization on the path usually contains values of the derivative of the path at some boundary points of S^*.

In the slightly general form of events $\{Z_s \geq \zeta(s)\ \forall s \in S\}$, where $\zeta \in R$ is any function belonging to the RKHS, the results of [135] on infinite intersections make it possible to determine the most probable paths to high buffer levels in priority queues exactly, instead of the heuristic approximations used in [133]. (Since the differences are numerically negligible, this is also an argument in favor of these heuristics.) However, the simple definition of the lower priority queue given above must be changed to the following. Let the service rate be c, and define first the cumulative capacity available for the lower priority traffic as $C_t^2 = \sup_{s \leq t}(cs - A_s^1)$, and then the lower priority queue as

$$Q_t^2 = \sup_{s \leq t}(A_t^2 - A_s^2 - (C_t^2 - C_s^2)) \tag{2.5}$$

(A^i denotes the cumulative input of class i). This definition agrees with the old one when the input processes are non-decreasing, but it is better in the Gaussian case since it makes the queue length process non-negative. The event of a large lower priority queue can be written as

$$\{Q_0^2 \geq x\} = \bigcup_{s \leq 0} \bigcup_{t \leq s} \bigcup_{a \geq x} \left(B_{s,t,a}^1 \cap B_{s,a}^2\right), \tag{2.6}$$

where $B_{s,t,a}^1 = \{A_u^1 - A_t^1 \geq c(u-t) + x - a\ \forall u \in [s,0]\}$, which is an infinite intersection event depending on A^1 only, and $B_{s,a}^2 = \{-A_s^2 = a\}$, which is a one-dimensional condition on A^2 only. This determines the possible shapes of the most probable path pairs, and it remains to optimize over the quantities s, t and a.

Delays can be analyzed with the same technique as queue lengths. The natural definition of the queueing delay in the lower priority class is

$$V_t^2 \doteq \inf\left\{u \geq 0 \,|\, Q_t^1 + Q_t^2 + (A_{t+u}^1 - A_t^1) < cu\right\}. \tag{2.7}$$

The structure "union of infinite intersections" is obtained once again:

$$\begin{aligned}\{V_0^2 \geq t\} \quad = \quad & \bigcup_{v \leq s \leq 0} \bigcup_{a \in \mathbb{R}} \Bigg(\bigcap_{u \in [0,t]} \{A^1(v,u) \geq -a + c(u-v)\} \\ & \cap \{A^2(s,0) = a\} \Bigg).\end{aligned}$$

Finally, a tandem queue allows a very similar analysis, see [136].

2.4.2 Delay Quantiles

Delay quantiles for the Gaussian voice traffic model are derived in [137]. Delay quantiles γ_k with $\Pr[\text{delay} > \gamma_k] = 10^{-k}$ play an important role when it comes to dimensioning and performance evaluation. Especially large links should be dimensioned such that delay quantiles remain relatively small. However, as related methods often suffer from time and memory limitations together with numerical instabilities, it is not necessarily simple to obtain numerical results for systems incorporating many streams. The superposition of a large number of homogeneous Markovian ON-OFF sources asymptotically approaches an Ornstein-Uhlenbeck Process (OUP) representing a Gaussian process with exponential autocorrelation function. For such an OUP/D/1 model, a closed-form expression is derived in [137] for delay quantiles as

$$\gamma_k \simeq \frac{\sigma_R}{\omega_R C m}\left(k\log(10) - \frac{7}{4}m - \left(\frac{k}{30}+\frac{1}{4}\right)m^2\right), \tag{2.8}$$

where $m = \frac{C-\mu_R}{\sigma_R}$. The parameter μ_R represents the mean rate, σ_R is the corresponding standard deviation, ω_R denotes the reciprocal time constant of the autocorrelation function of the rate, and C stands for the capacity of the link. By comparing, in the sense of the model, with exact results for a non-finite number of fluid flow on-off sources, we find that for the relevant parameter region defined by $\Pr[\text{source on}] \simeq 0.4$, $k \in [3, 6]$, $m \in [0, 1.6]$ and the number of sources $N > 100$, the related deviations of the approximated delay quantiles do not exceed 10%, which makes (2.8) a well-working approximation formula.

2.5 Processor Sharing Models

Processor Sharing (PS) models are widely applicable to situations in which different users receive a share of a scarce common system resource. In particular, over the past few decades PS models have found many applications in the field of the performance evaluation of computer-communication systems. The standard PS model consists of a single server assigning each customer a fraction $1/n$ of the service rate when there are n customers in the system. Cohen [48] generalizes the PS model to the so-called GPS model, where each customer receives a fraction $f(n)$ of the service speed when there are n customers at a node, where $f(\cdot)$ is, except for weak assumptions, an arbitrary function. The GPS model significantly enhances the modelling capabilities of the PS model. Interestingly, over the past few years the GPS model studied by Cohen in [48] in 1979 has received a renewed interest in the literature on performance of computer-communication networks (see, e.g., [138, 139, 140]).

A particularly attractive feature of (G)PS models is that in many applications they cover the main factors determining performance, and on the other hand, are still simple enough to be analytically tractable (see, e.g., the analysis in [141, 142, 48]). The present section gives a comprehensive overview of recent work on PS models within COST Action 279. Aside from the more generic analyses included in this section, we refer to [143, 144, 145, 146], discussed in chapter 4 on wireless networks, for specific applications of different PS models to analyze the performance of different mobile access technologies.

2.5.1 Sojourn Times for PS Models with Multiple Servers and Priority Queueing

In [61] is studied a PS model with multiple servers and two priority classes (without loss of generality). When the number of high-priority customers does not exceed the number of servers, C, each high-priority customer occupies a single server and is served at unit rate. When the number of high-priority customers is larger than C, the system switches to a processor sharing mode and the total service capacity C is equally shared among the high-priority customers. The service process of low-priority customers proceeds in a similar way, but with two specific restrictions: (i) high-priority customers have strict priority, in a preemptive resume fashion, over low-priority customers, and (ii) at any moment in time the only servers available to low-priority customers are those servers that are not used by high-priority customers at that moment.

In [61], the mean sojourn times in the multiserver queueing model with PS service discipline and two priority classes described above are studied. For the high-priority class, closed-form expressions for the mean sojourn times are presented in a general parameter setting, based on known results for the GPS model (see [48]). For low-priority customers, closed-form expressions are derived for several special cases: the single-server case where the service times of the low-priority customers are exponentially distributed, and the multiple-server case with exponential service times with the same means. In all other cases, exact explicit expressions for the mean sojourn times of the low-priority customers cannot be obtained. Therefore, a simple and explicit approximation is proposed and tested. Numerical results demonstrate that the approximation is accurate for a broad range of parameter settings. As a by-product, it is observed that the mean sojourn times of the low-priority customers tend to decrease when the variability of the service times of the low-priority customers increases.

Application of these results to the setting with elastic Transmission Control

Protocol (TCP) traffic is given in [59], discussed in Section 1.5.3 of the chapter on IP-based networks of this report.

2.5.2 Throughput Measures for PS Models

In [147] are specified, derived, and compared a set of throughput measures in PS queueing systems modeling a network link carrying elastic TCP data calls, e.g., file downloads. The available service capacity is either fixed, corresponding to a stand-alone dedicated General Packet Radio Service (GPRS) network, or randomly varying, corresponding to an integrated services network where the elastic calls utilize the capacity left idle by prioritized stream traffic such as speech.

While from the customer's perspective the *call-average throughput* is the most relevant throughput measure, in PS systems this quantity may be hard to determine analytically, and this is an important reason to assess the closeness of a number of other throughput measures. Alternatives applied to approximate the call-average throughput are the *time-average throughput* [148, 149, 150], defined as the expected throughput the "server" provides to an elastic call at an arbitrary (non-idle) time instant, and the *ratio* of the expected transfer volume and the expected sojourn time [58, 151, 152, 62].

In [147] is introduced a new throughput measure that can be analyzed relatively easily, the *expected instantaneous throughput*, i.e., the throughput an admitted call experiences immediately upon admission to the system. The experiments demonstrate that the newly proposed expected instantaneous throughput measure is the *only* one among these throughput measures that excellently approximates the call-average throughput for each of the investigated PS models over the entire range of elastic traffic loads. In particular for the model integrating speech and data traffic, the other throughput measures, such as the time-average throughput or the ratio of the expected call size and the expected sojourn time, significantly underestimate the call-average throughput. An intuitive reasoning for the generally near-perfect fit of the expected instantaneous throughput is that, apparently, the throughput an elastic call experiences immediately upon arrival is an excellent predictor of what the call is likely to experience throughout its lifetime. Moreover, among the considered throughput measures, the expected instantaneous throughput is the *only* approximate measure that is truly *call*-centric.

The numerical experiments further reveal that the expected call-average throughput of elastic calls in the considered PS models is to a considerable degree *insensitive* to both the variability of the available capacity and the call duration distribution. This insensitivity does not hold if the data performance is measured by the expected sojourn time.

2.5.3 PS Models with State Dependent Blocking Probability and Capacity

In [56], PS models with state dependent blocking probability and capacity are investigated. It is shown that if the blocking probability and capacity of a PS system are functions of the system state, the state probabilities are insensitive to the detailed distributions of thinking time and file size. This insensitivity property allows the evaluation of performance through the offered traffic only. From here, formulas to calculate state probabilities, blocking probabilities, and the conditional mean sojourn times of users are derived by using the so-called A-formula. By proper transformation of the state probabilities to GPS rates it is possible to calculate the resources obtained by every source via the use of the convolution algorithm. This technique allows the evaluation of the performance of sources having individual bandwidth (multi-rate traffic) under the GPS strategy.

2.6 Multilevel Processor Sharing Models

Multilevel Processor Sharing (MLPS) scheduling disciplines, introduced in [49] permit to model a wide variety of non-anticipating scheduling disciplines. A discipline is *non-anticipating* when the scheduler does not know the remaining service time of jobs. Such disciplines have recently attracted attention in the context of the Internet as an appropriate flow-level model for the bandwidth sharing obtained when priority is given to short TCP connections [153, 154, 155, 156].

An MLPS scheduling discipline is defined by a finite set of level thresholds $a_1 < \cdots < a_N$. A job belongs to level n if its attained service is at least a_{n-1} but less than a_n. Between these levels, a strict priority discipline is applied, with the lowest level having the highest priority. Within each level n, an internal discipline is applied, belonging to the set $\{\mathrm{FB}, \mathrm{PS}, \mathrm{FCFS}\}$. The foreground-background (FB) discipline is also known as least-attained-service (LAS), giving priority to the job with the least attained service.

In [157] and [158], the mean delay of those MLPS disciplines whose internal disciplines belong to the set $\{\mathrm{FB}, \mathrm{PS}\}$ is compared to that of the ordinary PS discipline. In [157], it is proved that such MLPS disciplines with just *two* levels are better than PS with respect to the mean delay, whenever the service time distribution is of type Decreasing Hazard Rate (DHR), e.g., hyperexponential. In [158], a similar result is derived for such MLPS disciplines with *any* number of levels.

In [159], the mean delay is compared among all MLPS disciplines. The

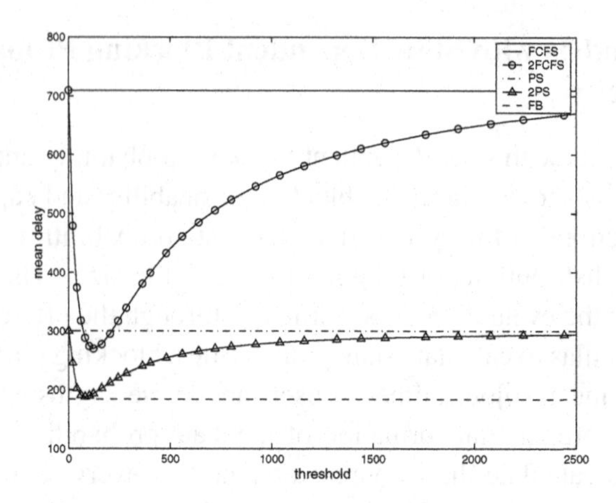

Figure 2.8: Mean delay as a function of the level threshold a for disciplines $2\text{FCFS}(a)$ and $2\text{PS}(a)$. The three horizontal lines correspond to the mean delay of disciplines FCFS, PS, and FB

main result states that given an MLPS discipline, the mean delay is reduced, under the DHR condition, if a level is added by splitting an existing one. Furthermore, it is shown that given an MLPS discipline, the mean delay is reduced, under the DHR condition, if an internal discipline is changed from FCFS to PS, or from PS to FB. These two results define a natural partial order among the MLPS disciplines: if one MLPS discipline is derived from another by splitting levels and/or changing internal disciplines from FCFS to PS, or from PS to FB, then the mean delay is reduced under the DHR condition. This reduction is illustrated in Figure 2.8, where the mean delay is depicted as a function of the threshold a for two-level disciplines $2\text{FCFS}(a)$ and $2\text{PS}(a)$ and a Pareto service time distribution. As stated, the mean delay for any $2\text{PS}(a)$ discipline is less than that of the corresponding $2\text{FCFS}(a)$ discipline or the ordinary PS discipline. The mean delay of an MLPS discipline with just a few levels gets close to the minimum feasible delay, achieved by FB. Since flow size distributions in the Internet typically satisfy the DHR condition, these results are interesting in view of recent work that proposes to provide differential treatment to flows on the Internet based in just two classes: mice and elephants.

2.7 Other Continuous-Time Queueing Models

This section discusses a variety of continuous-time queueing models and associated analysis techniques studied in COST Action 279 .

2.7.1 Instantaneous and Averaged Queue Length in an M/M/1/K Queue

In [160] is studied the dynamics of the joint process of the instantaneous queue length $L(t)$ of an M/M/1/K system together with the exponentially averaged queue length $S(t) = \int_0^\infty L(t-u)\alpha e^{-\alpha u} du$, where α is a weighing parameter. The arrival and service rates are denoted by λ and μ. The state of the system is specified by the pair $(L(t), S(t))$, i.e., one discrete and one continuous variable. The setting is very similar to that of a fluid queue driven by an MMRP.

The evolution of the joint distribution of the state variables is governed by a system of coupled ordinary differential equations (ODE) for the partial cumulative distribution functions $F_i(t, x) = \Pr[L(t) = i, S(t) \leq x]$,

$$\frac{\partial}{\partial t}F_i(t,x) - \alpha(x-i)\frac{\partial}{\partial x}F_i(t,x) = (\lambda F_{i-1}(t,x) - \mu F_i(t,x))1_{i>0}$$
$$+(\mu F_{i+1}(t,x) - \lambda F_i(t,x))1_{i<K}, \quad i = 0, \ldots, K. \tag{2.9}$$

An analytical stationary solution to these equations is found in a few special cases. A general stationary solution is not known and is believed not to have a simple form. Therefore, different alternative approximate ways for obtaining the stationary distribution are developed in [160].

Two of the methods consider the temporal behavior of the state distribution. By Kolmogorov's theorem, starting from any initial distribution the system will eventually approach an equilibrium, i.e., integrating the equations in time is inherently stable. The first of the methods considers the evolution of the system in continuous time, whereas in the second approach an embedded system in discrete time is studied. A disadvantage of these methods is that the equilibrium is only approached asymptotically, and with a long averaging time the convergence is slow. The third method focuses directly on the equilibrium distribution but using an approximation. The method applies the stochastic discretization approach, where the deterministic evolution of the continuous variable is replaced by small stochastic transitions, thus allowing the use of standard methods of Markovian systems.

2.7.2 M/D/1/K Vacation Queue

In [16], a queueing model is adopted where voice packets are fed into a finite buffer and served by a server representing the output link. The aggregate voice traffic is modelled by a Poisson process. Due to the presence of best-effort traffic and to the DiffServ-compliant non-preemptive priority scheduling, the operation of the server is considered in an exhaustive service and multiple vacation scenario. That is, the server serves voice packets until the buffer becomes empty. At the finishing instant of the service, if the server finds the queue empty, it takes a vacation. If there are still no voice packets in the queue when the server returns from its vacation, it takes another vacation, and so on. The vacations of the server correspond to the situation where the output link is occupied by the best-effort traffic. The assumption of multiple vacations implies that the offered load of the best-effort traffic is sufficiently high to utilize immediately the link capacity whenever no voice packet is present. The vacation time is assumed to be the time needed for transmission of a best-effort packet with maximum transmission unit (MTU) size. In effect, a finite M/D/1/K queue with exhaustive service and multiple server vacations is obtained.

The steady-state solution of this queueing model is obtained, and useful quantities concerning packet loss probability and arbitrary percentile of delay are derived. The latter quantity is particularly valuable, because it stands for a statistical upper bound on the jitter, which is the main factor leveraging the perceived quality of voice connections.

2.7.3 BMAP/G/1 Queue with Feedback

Message transmission in wireless communication networks includes procedures for error correction at several layers of the protocol stack, e.g., at the data link layer (DLL) and transport layer. These procedures perform the repeated transmission of those protocol data units (PDU) that were transmitted with errors. An adequate performance model of such procedures is determined by specific feedback queues.

In [161], matrix-geometric modelling techniques are used to analyze and calculate the performance characteristics of such a telecommunication channel where the probability of a corrupted transmission is fluctuating. Such situations typically occur during information transmission in mobile networks when the users cross cell boundaries and the interference conditions change drastically.

First, the transmission process is modeled in terms of the BMAP/G/1 queue with feedback where the behavior of the input and the error probability depend on the state of a Markovian synchronous random environment. The latter describes the random changes of the interference conditions determining the success of a service completion. Here the probability of a repeated service, which can be interpreted as the error probability of a transmitted PDU, can change according to the state of that random environment. A general Batch Markovian Arrival Process (BMAP) describes the arrival stream of customers, modelling the batch arrival of radio blocks of a segmented message at the DLL. Applying the machinery of matrix-geometric methods, the resulting model is characterized by a discrete-time Markov chain with quasi block-Toeplitz structure embedded upon service completions. Then necessary and sufficient conditions for the existence of the corresponding steady-state distribution of the queue length at these embedded epochs are determined. Furthermore, the latter is characterized as the unique solution of a generalized variant of the Pollaczek-Khintchine equation using a generating-function approach. Finally, the stationary queue-length distribution at arbitrary epochs is determined and an algorithm for its calculation is sketched.

The model of [161] includes as special cases both Takacs' single-server feedback and a BMAP/G/1 queue operating in a synchronous random environment without feedback.

2.7.4 Two-Class Non-Preemptive Priority Queue

A non-preemptive priority scheduler with two traffic classes and a separate buffer for each class is analyzed in [15]. The arrival streams of each class are assumed to be independent Poisson processes. The packet sizes are generally distributed. The aim of the analysis is to determine the maximum admissible low-priority traffic load under the assumption that the buffer size is finite and a constraint on the packet loss ratio is given. As a first step of the analysis, the system with two priority classes is translated into an equivalent system with a single queue and a single server where only low-priority traffic is present. For this system, the impact of the high-priority traffic is accounted by extending the service times experienced by low-priority packets. Furthermore, the system is analyzed using the diffusion approximation method, which only requires the two first moments of the packet arrival and departure processes. Since the input is Poisson, the crucial point is the departure process, characterized by the mean and the variance of the extended packet service times. These parameters are obtained from the busy period analysis of the high-priority traffic exploiting a functional equation approach. Finally, the probability distribution function of the low-priority queue size is determined and the admissible load is calculated. Numerical results illustrate the impact of the traffic load and the packet sizes of both priority classes on the admissible load. A comparison with the results obtained from the reduced service rate (RSR) method [162] is also included. The results indicate that when the high-priority traffic load is light or the high-priority packets are small compared to the low-priority packets, the impact of the high-priority traffic can be sufficiently captured by the first moment as in the RSR method. However, in the remaining cases the diffusion approximation method provides more accurate results than the RSR method.

2.8 Queueing Networks

The previous sections of this chapter deal with isolated queueing systems. In order to assess the performance of a communication network, however, it is also necessary to study networks of queues. This section is devoted to the COST 279 work on queueing networks. First, an approximate method to calculate end-to-end delay characteristics is presented. Next, a technique to determine the evolution of the characteristics of a traffic stream when it proceeds through a network is discussed. The latter can be useful to derive more accurate end-to-end performance characteristics.

2.8.1 End-to-End Delay Characteristics

The end-to-end delay is an important QoS parameter for real-time services. In [163], an analytical model to calculate end-to-end delays in packet networks is considered. The aim is to calculate the distribution of the end-to-end delay for a particular path consisting of a series of nodes. It is assumed that all the waiting times in the nodes in the end-to-end path are *statistically independent*; this is a key assumption to obtain the end-to-end delay by convolution. For queueing networks with FCFS queueing discipline this property only holds for the acyclic form of Jackson Networks, where a packet visits a node at most once. In [164], however, it is argued that if the load from a particular flow is only a small fraction of the total amount of traffic at a node and the input processes to the network are "smoother than Poisson," i.e., with less variability, then the independence assumption will be quite reasonable and will represent a worst case scenario. Therefore, the M/G/1 queue is taken as the model to find the waiting time distribution in each node, and then the convolution is applied to obtain the end-to-end waiting time distribution.

If all nodes have identically distributed service times, the corresponding convolution may be substantially simplified, and closed-form expressions are obtained in terms of derivatives with respect to the load parameter. Special emphasis is put in [163] on the case with constant service times, since this is an important case for applications. Numerical results show that end-to-end delays in chains for up to 20 nodes may be analyzed without numerical difficulties. It is also possible to extend some of the results to cover convolutions between equally loaded groups of queues with different service time distributions in each group.

2.8.2 Evolution of Traffic Characteristics

The evolution of the characteristics of the interarrival and interdeparture times between voice packets as they proceed through a number of network nodes is studied in [165]. Each network node is assumed to have an infinite-capacity buffer. The arrival process in a node is modelled as the superposition of a single tagged voice stream and an independent background process that aggregates the remaining traffic sources. Since the load of a single voice stream is very low compared to the load of the aggregate traffic, the tagged voice packets can be represented as *markers*, i.e., packets of size zero. At the entrance of each network node, one thus has the tagged marker stream and the background stream. The tagged marker stream is characterized by the interarrival times between successive markers, which are assumed to be identically distributed but may be dependent. The background arrival process is described on a slot-per-slot basis according to a general iid process, independent of the tagged marker stream.

In [165], first, an expression for the PGF of the interdeparture times of the voice packets after one stage is established. The PGF of this interdeparture time is then used as the PGF of the interarrival times of the voice packets in the next stage, in order to assess the evolution of the interarrival-time characteristics throughout the network. Following, the PGF of the interdeparture times between three successive voice packets (in case two successive interarrival times may be dependent of each other) is also calculated.

2.9 Models for Optical Buffers and Networks

Optical packet switching (OPS) and optical burst switching (OBS) seem promising techniques to cope with the explosive growth of the Internet traffic. This section presents some new models and analysis techniques to evaluate the performance of burstification queues, optical cross-connects, and various types of fiber delay line buffers. We refer to chapter 5 on optical networks for a further discussion of the obtained results.

2.9.1 Burstification Queues

In the edge routers of an OBS network, IP packets are assembled into bursts. Core OBS routers forward these bursts in the optical domain through the OBS network. An OBS edge router can be decomposed into multiple burstification units (BU). Each BU consists of a set of separate output queues. In [166], the burstification of a single isolated output burst queue is investigated. First, a single-threshold burst assembly mechanism is studied, where bursts are released whenever they contain exactly S packets. In this case, especially for low throughputs, the packets may have long delays. Therefore, as a next step, also a two-threshold model is investigated, where besides a threshold on size, a threshold on a burst's age is imposed. Thus, bursts are also released if, since the start of their assembly, a time T has expired. For both queueing models, some performance characteristics are calculated. Results include the Probability Mass Function (PMF) of the system content in the output queue of the OBS edge routers, the PMF of the delay of the bursts, defined as the interdeparture time between two bursts, and the PMF of the delay of the individual packets. Using these PMFs, the mean values, variances, and tail distributions of the system content and of the burst and packet delays are derived.

2.9.2 Optical Cross-Connects

An asynchronous bufferless optical cross-connect using a shared wavelength converter pool with sharing on a per-output-link basis is studied in [167]. In the model studied, an incoming optical burst, or optical packet, is blocked either because there is no available wavelength on the output link, or the incoming burst requires conversion but the converter pool is fully occupied. The goal is to exactly calculate the steady-state blocking probabilities as a function of the basic system parameters, e.g., mean arrival rate, arrival statistics, and converter pool size. Using the traditional model of Poisson burst arrivals, exponential burst lengths, and uniformly distributed burst colors, this problem is formulated in [167] as one of finding the steady-state solution of a finite Continuous-Time Markov Chain (CTMC) with a block tridiagonal infinitesimal generator or, equivalently, that of a finite non-homogeneous QBD process. The number of converters in use form the phase of the QBD process, whereas the level process is dictated by the number of wavelengths in use. Although matrix-geometric forms of solutions are available for finite and infinite QBDs with a homogeneous structure, i.e., where block rows repeat, numerical studies of non-homogeneous QBDs are rather rare. A stable and numerically efficient technique based on block tridiagonal LU factorizations is proposed for exactly calculating the steady-state probabilities of the non-homogeneous QBD. It is shown through numerical examples that blocking probabilities can exactly and efficiently be found even for very large systems and rare blocking probabilities. The formulation of the problem is also extended to phase-type (PH-type) burst arrivals by incorporating the phase of the arrival process in the phase process of the non-homogeneous QBD using Kronecker calculus. The results obtained using PH-type arrivals clearly demonstrate that the coefficient of variation of burst interarrival times is critical in burst blocking performance, and therefore burst shaping at the edge of the burst/packet switching domain can be used as a proactive congestion control mechanism in next-generation optical networks.

2.9.3 Fiber Delay Line Buffers

In the design of all-optical switches, the lack of optical Random Access Memory (RAM) poses a big challenge. Besides wavelength conversion and deflection routing, the use of fiber delay lines (FDL) can help alleviate the output port contention problem. These FDLs are passive components that can delay an optical packet or an optical data burst for a fixed time. Usually, an FDL buffer implements the delays $0 \cdot D, 1 \cdot D, \ldots, N \cdot D$, where D is the so-called *granularity* and $N \cdot D$ can be considered as the *capacity* of the FDL buffer. Note that not all delays can be thus obtained, typically leading to the creation of voids in the scheduling and to underutilization of the output channel. For this reason, FDL buffers are also sometimes called degenerate buffers.

The performance of a single-wavelength FDL buffer is analyzed in [168], for the synchronous case, and in [169], for the asynchronous case. The quantity of interest in the analysis is the so-called *scheduling horizon*. It is defined as the earliest time at which the channel will become available again, and can be considered the equivalent of the unfinished work in non-degenerate buffers. If one denotes by H_k this scheduling horizon as seen by the k-th arrival, one can easily establish, assuming an infinite FDL buffer, the following recursion:

$$H_{k+1} = \left[B_k + D \left\lceil \frac{H_k}{D} \right\rceil - \tau_k \right]^+ . \tag{2.10}$$

Here B_k denotes the size of the k-th burst, and τ_k the interarrival time between the k-th and $(k+1)$-th burst. One easily recognizes part of the evolution equation for non-degenerate buffers, involving the operation

$$[\ldots - \tau_k]^+ . \tag{2.11}$$

Under the usual assumptions of iid interarrival times and iid. burst sizes, the solution to this problem in the transform domain is well-known. The part

$$D \left\lceil \frac{H_k}{D} \right\rceil \tag{2.12}$$

reflects the finite granularity of the FDLs. In discrete time, as was done in [168], this operation on random variables can be translated into an operation on their PGFs, by using an identity involving the complex D-th roots of unity. By combining both partial solutions, one obtains in the end the PGF of the scheduling horizon H in equilibrium. From the latter, one can obtain several measures of interest, such as the maximum tolerable load, i.e., the load at which the infinite system becomes unstable. Due to the creation of voids, this load is typically less than unity; it also shows a slight dependency on the burst size distribution. Further, a heuristic can be used to map the tail probabilities $\Pr[H > M \cdot D]$ to loss probabilities in a finite system of capacity $M \cdot D$. An optimum granularity D_{opt} exists, establishing a compromise between increasing capacity ($D \to \infty$) and small voids ($D \to 0$). This optimal value not only depends on the burst size distribution, but also on the load offered to the system. Results for the asynchronous case were obtained in [169] by taking the appropriate limits for slot lengths going to zero. A more direct analysis, explicitly taking into account the continuous-time nature of asynchronous FDL buffers, leads to the same results. In the end, one obtains the Laplace-Stieltjes transform of the scheduling horizon in equilibrium, from which other results can be obtained, proceeding as in the discrete-time case. Here too, the optimal granularity is sensitive to both the load and the burst size distribution.

In [170], the performance of an optical packet switch is investigated. Specifically, an FDL-structure consisting of N delay lines with increasing lengths is considered. It is assumed that the optical packets have deterministic lengths and the i-th delay line ($i = 1, .., N$) has a length of i times the packet length. Time is slotted, where one slot corresponds to the time needed to transmit a packet. The numbers of packet arrivals are assumed to be iid from slot to slot. The scheduling discipline is smallest FDL first: if i packets arrive at the same time, they are put in the i delay lines with the smallest lengths, with $i - N$ packets lost if $i > N$. First, an expression for the steady-state PGF of the delay line content of the FDL-structure with increasing lengths is derived. From this PGF, the Packet Loss Ratio (PLR) in an output buffering optical packet switch is then calculated. Through some figures, the impact of the number of delay lines and the load on the PLR is shown. An important conclusion is that putting one or two delay lines at each output can reduce the PLR significantly. Adding even more delay lines per output does not reduce the PLR significantly, though. If a lower PLR has to be obtained, the scheduling discipline discussed in [170] is not sufficient, and more complex scheduling methods are necessary.

The performance of a two-stage optical buffer is investigated in [171]. Fixed-length packets enter the first stage according to the simple routing scheme investigated in [170]. This scheme routes incoming packets to delay lines such that there is no contention at the input of the delay lines. However, contention at the output is possible. The output traffic is then routed to the FDLs of the second stage, again according to the simple routing scheme. Using a generating-functions approach, the PLR of the buffer structure under consideration is obtained.

Chapter 3
Traffic Measurement, Characterization, and Modeling

Markus Fiedler
Blekinge Institute of Technology, Karlskrona, Sweden

Contributors:
Patrik Arlos (Blekinge Institute of Technology, Sweden), Hans van den Berg (TNO Telecom, Netherlands), Mine Çaglar (Koç University, Turkey), Markus Fiedler (Blekinge Institute of Technology, Sweden), Guoqiang Hu (University of Stuttgart, Germany), Jorma Kilpi (VTT Information Technology, Finland), Udo Krieger (Otto Friedrich University Bamberg, Germany), Petteri Mannersalo (VTT Information Technology, Finland), Sándor Molnár (Budapest University of Technology and Economics, Hungary), Luca Muscariello (Politecnico di Torino, Italy), Philippe Olivier (France Telecom, France), Öznur Özkasap (Koç University, Turkey), Adrian Popescu (Blekinge Institute of Technology, Sweden), Marie-Ange Remiche (Université Libre de Bruxelles, Belgium), Kavé Salamatian (Université Paris VI, France), Paulo Salvador (Telecommunications Institute, Portugal), Kurt Tutschku (University of Würzburg, Germany), Sabine Wittevrongel (Ghent University, Belgium).

3.1 Introduction

Pioneering work revealing yet unknown properties of network traffic and their implications has stipulated research during the recent two decades. It became obvious that the well-established teletraffic theory that had served for telephony systems for many decades was not able to capture essential aspects of traffic behavior in data networks [172]. In the context of research on Broadband Integrated Services Data Network (B-ISDN) systems, the bursty nature of data traffic gained much attention during the 1980's and led to the establisment of new traffic models. In the early 1990's, Internet traffic attracted the researcher's interest. Especially the discovery of yet unknown scaling behavior [173, 174] derived from measurements boosted work within the scope of this chapter. Since then, the famous Bellcore traces [175] from 1989 have served as basis for many investigations.

Traffic measurements and their interpretation constitute a *descriptive ap-*

J. Brazio et al. (eds.), Analysis and Design of Advanced Multiservice Networks Supporting Mobility, Multimedia, and Internetworking, 85–113.

proach of what is going on in the "black box" network [176]. As they help to discover the properties of certain network traffic and its interaction with other traffic and network resources (cf. [177]), they play an important role for network design and control, and of course for related research and development. In order to deliver relevant data with regard to the target of a measurement, rather technical issues related to measurement methods and quality of measurements have to be addressed. Thus, Section 3.2 reports activities and work on *traffic measurement techniques* within COST Action 279.

Pure measurement data alone without interpretation can at best serve as historical data [176] documenting the evolution of network traffic. A first important step in interpreting measurements is *traffic characterization*, which aims at providing a rather qualitative description of traffic behavior and peculiarities. Work in the scope of COST 279 on this issue is addressed in Section 3.3.

Even though traffic characterization based on measurements can give hints on reasons for certain traffic patterns, it does not necessarily allow for so-called "as-if" studies, e.g., yielded by variations of parameter settings. A *constructive approach* [176] is devised, aiming at models producing the same output as the system under study given similar conditions. The modeling phase is usually followed by a resolution phase incorporating simulation or analytical analysis. Thus, *traffic modeling* is about a quantitative description of traffic and device properties, which implies the needs for model parameter matching, e.g., using the EM method [176, 178], and model validation, e.g., through measurements [177]. Traffic modeling has a long tradition within the series of COST Actions leading to COST 279, and Section 3.4 comprises a comprehensive overview of recent activities and results.

In any case, outputs from both measurements and models have to be interpreted in order to relate effects to causes. Moreover, measurement results may be "consumed" [179] instantaneously, e.g., by a network monitoring device or some control entity. Section 3.5 presents a framework for measurement interpretation and examples of use, such as traffic estimation, classification, and control.

3.2 Traffic Measurement Techniques

The following subsections provide an overview of tools used in various studies and discuss achievements in measurement methodology within the scope of COST Action 279.

3.2.1 Measurements Principles

Passive Measurements

The following definition is given in [180]: "Passive measurements are by nature non-intrusive. In this class of measurements, traffic parameters are monitored at a particular point of the network such as a router or a Point of Presence (POP)." By monitoring, i.e., basically counting or collecting traffic in different places at the same time, the analyst may get a holistic view on what is going on in a network and collect statistical data to build models upon.

The most popular basis for traffic characterization and modeling is the collection of *traffic traces*, i.e., sequences or parts of sequences of IP packets or Ethernet frames together with corresponding time stamps. Depending on the scope of the corresponding study, header information belonging to different protocol layers (link, network, transport, and application) and in some cases parts of the payload or even the whole payload are collected. Especially in the latter case, privacy issues may have to be observed. As the collection of traces plays a central role in measurement-related activities and requires special technical considerations, it is discussed in detail in the separate Subsection 3.2.2.

Network Monitoring

For network operation and management purpose there exists a widely deployed infrastructure for passively monitoring performance-related data. Almost each non-low-cost IP-enabled device offers an agent that can be polled via the Simple Network Management Protocol (SNMP). If available, the Remote Monitoring (RMON) extension delivers additional performance data, e.g., link utilization history and packet size statistics. For network monitoring purposes, the polling is done on rather large time scales, typically 5 to 60 minutes. As advanced traffic load modeling needs information about the traffic on timescales of seconds and less, [181] investigates to what extent SNMP agents can be used for such a purpose. Depending on the device under test, counter update intervals of several seconds or response times of up to half a second were observed. This illustrates the need for evaluating the response behavior of SNMP agents before using them for quantitative measurements on short time scales.

Successful efforts to infer traffic behavior on short time scales on backbone links from medium-term measurements, on the order of 15-minute intervals, are presented in [57], cf. Section 1.5.3.

Active Measurements

According to [180], "Active measurements are more intrusive, as they inject traffic into the network. The rationale behind active measurements is that estimating the end-to-end Quality of Service (QoS), as sensed by a real application, can only be done by putting oneself in place of the real application." Thus, active measurements reflect the experience of a network user rather than a holistic view of the overall performance of a network.

The simplest and most widely known active measurement tool is `ping`, which allows for the estimations of Round-Trip Time (RTT) and loss quotas. In [182] is presented a further development for wireless networks called *Wave-Ping* (`WVPing`). When invoked, this tool reads status information from the wireless network interface card driver. Thus, the important connection between RTT and signal quality, noise level, Signal-to-Noise Ratio (SNR), and number of erroneous packets at the Medium Access Control (MAC) level can be displayed. Furthermore, monitoring the identity of the radio channel used reveals handovers. `WVPing` results are shown in Section 3.3.6.

Another example of an active measurement is given by the *packet-pair technique* mentioned in [180]. Two packets of length L are sent into the network in a back-to-back fashion. If they arrive dispersed at the receiver, i.e., spaced by time τ, then they have experienced a bottleneck of capacity $\mu = L/\tau$. Analogous to passive measurements, these packets have to be collected and time-stamped at the receiver side, cf. Subsection 3.2.2.

Hybrid Method

In [183] is presented a bottleneck characterization method based on passive measurements. The results of those measurements have to be compared on-line in order to draw conclusions about the nature of the bottleneck. This implies the need for transmitting measurement results between sender and receiver, which can be considered as an active component, since it consumes network resources. The method can also be used to evaluate the perception of probing traffic as applied by active measurements as described above. However, due to the possibility to compare measurement results, the method can take real sending patterns into account, e.g., if packets are not or cannot be sent back-to-back [180].

3.2.2 Traffic Trace Collection

Tools and Setups

A very common tool for collecting IP packets is `tcpdump` [184], which is used for instance in [185, 186, 187, 188, 183, 189, 190, 191, 192, 177, 193]. There exists also a Windows-based variant called `windump` [194], used for instance in [193].

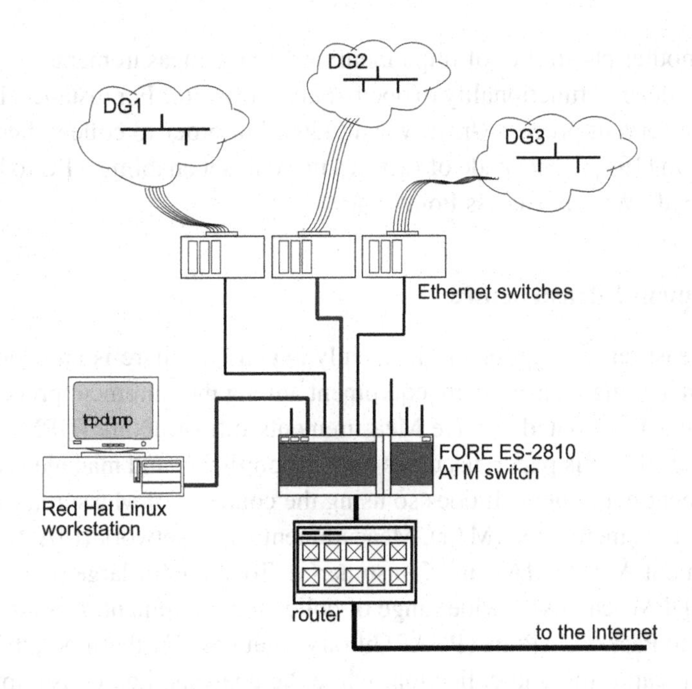

Figure 3.1: Example of the use of `tcpdump` on a monitoring port of a switch [187]

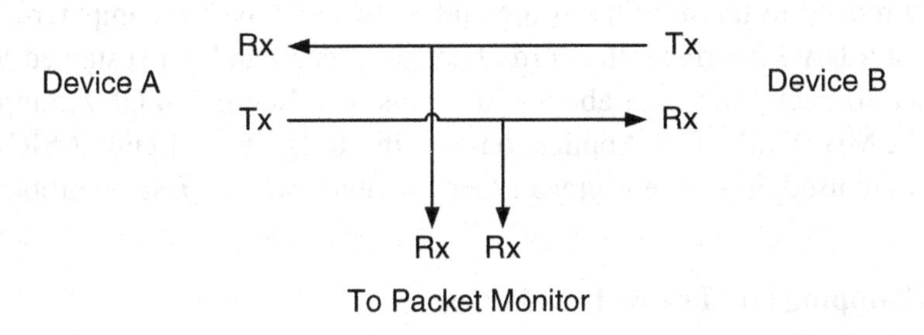

Figure 3.2: Cabling diagram of a wiretap

Usually the tool runs on a sender, receiver, or dedicated measurement computer. The latter may be connected to a monitoring port of a switch [187], as in Figure 3.1, or to a wiretap [192, 193], as in Figure 3.2. In wireless scenarios, the measurement computer listens to Wireless Local Area Network (WLAN) traffic; corresponding setups are described in [195, 177].

`Tcpdump` traces have to be post-analyzed to yield the desired information. For example, the tool `Tstat` [190, 192, 177] allows for the reconstruction of TCP sessions. In [191] is described how to identify signalling and download packets for the popular peer-to-peer service eDonkey [196].

Yet another possibility of implementing passive measurements consists in adding the desired functionality to open-source software. For instance, in [197] a Gnutella servent process `Gnut` was modified in order to collect the desired statistics. In [195] use is made of features in a chipset enabling a PC to become a home-made WLAN Access Point.

Measurement Infrastructure

As the measurement equipment is usually expensive, there is an economical gain when sharing measurement equipment among measurement processes. A concept of a Distributed Passive Measurements Infrastructure (DPMI) is presented in [179]. This paper discusses how to coordinate and manage such joint measurement equipment. It does so using the concepts of Measurement Point (MP), Measurement Area (MAr), Measurement Area Network (MArN), Super Measurement Area (SMA), and Consumer(s). To allow for large scale deployment the DPMI can use a wide range of collection equipment, from dedicated hardware to Packet CAPture (PCAP) library solutions. Furthermore, the DPMI focuses on capturing and collection, while the analysis, i.e., consumption, of the data is up to the user. Figure 3.3 presents an overview of the key components of the DPMI. The performance obtained by the DPMI is obviously directly related to the used hardware and software. Given a computer-based MP with a DAG interface, the current software can handle a sustained load, i.e., back-to-back frames, of about 850 Mbit/s on a Xenon-2.4 GHz computer with 512 MB RAM. If an Application-Specific Integrated Circuit (ASIC) solution were used, link speeds greater than 10 Gbit/s should pose no problems.

Time Stamping and Frame Loss Issues

In certain cases, e.g., if one-way delays are to be measured and modeled, a synchronization of the time stamping at different locations is important [179]. Such synchronization can be obtained via the Network Time Protocol (NTP) or the Global Positioning System (GPS).

In any case, precise time stamping of the packets is a crucial issue for drawing correct inferences about the time behavior of traffic. The precision of time stamping, however, is recognized as a critical issue [180]. On this background, in [193] is compared the time stamping performance of the special measurement card DAG [198] and the tools `tcpdump` [184] and `windump` [194]. The scenario is shown in Figure 3.4, in which a common signal, generated by the "Source" computer, is used to evaluate in parallel three measurement systems. The wiretaps replicate the 100 Mbit/s full-duplex Ethernet signals onto

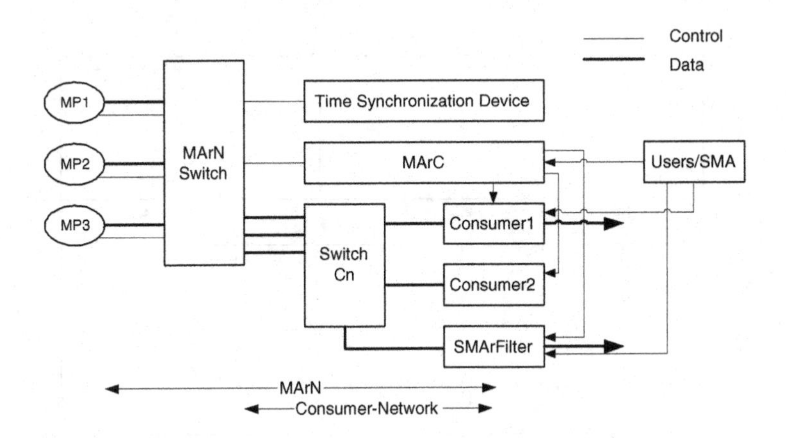

Figure 3.3: DPMI components (description found in the text)

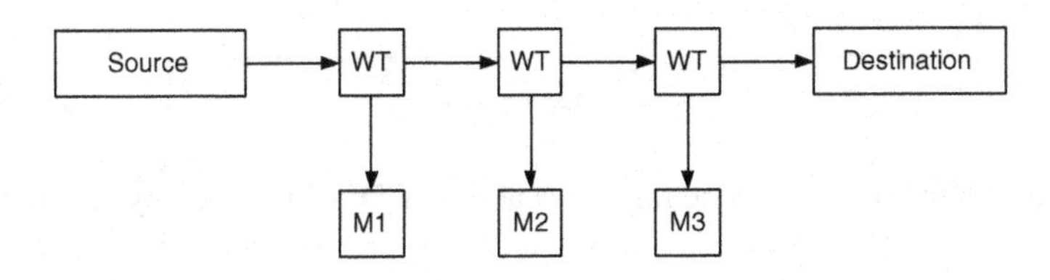

Figure 3.4: Setup for time stamp comparison (WT = wiretap, M = measurement computer) [193]

the monitoring ports without interference. Measurement point M1 contains a DAG card, M2 runs `tcpdump`, and M3 runs `windump`.

A DAG card performs the time stamping upon reception of the very frame at a time resolution in the order of 100 ns. `Tcpdump` and `windump` assign the time stamps once the corresponding kernels "discover" a packet, which implies some kind of uncertainty about when the corresponding frame *actually* passed by on the link. Another risk is that of data loss, i.e., unregistered packets. Obviously, missing packets may turn a trace worthless. DAG is able to stamp frames at line speed, but they have to be consumed instantaneously in order to avoid buffer overflow. The software-based time stamping tools may drop packets as well, but such events are usually reported. However, in [193] is revealed a case when `windump` was not even able to count all incoming packets. Regarding the quality of the time stamping as such, a typical case of a complementary distribution function of the frame inter-arrival time is shown in Figure 3.5. In this case, the frame size was 424 Bytes, corresponding to a theoretical frame rate of 29481 fps. The Linux packet generator in

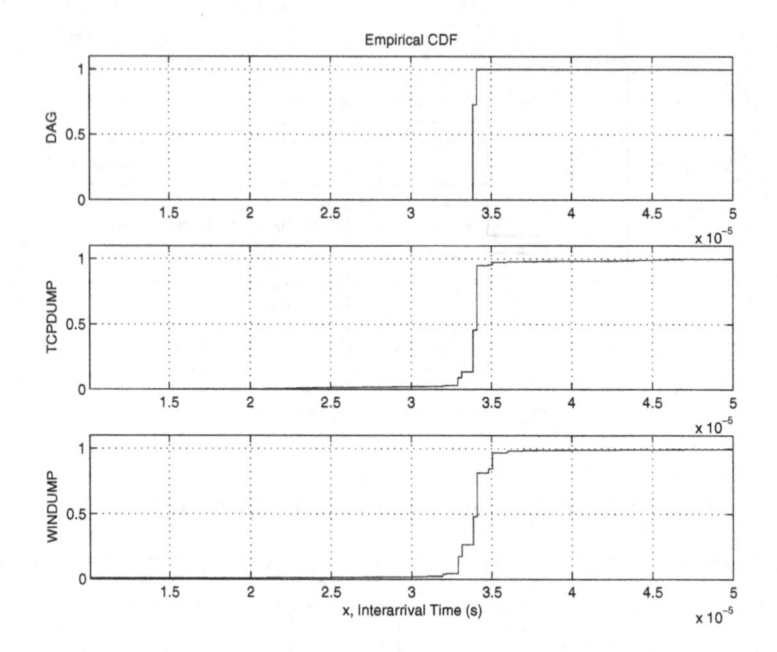

Figure 3.5: Comparison of frame inter-arrival time CDFs measured by different tools

"Source" yielded 29477 fps, which means that frames are sent almost back-to-back. Computers M2 and M3 had rather modest computing power. Obviously, both tools do not reach the same degree of sharpness as the reference time stamping by DAG. Multimodal distributions and distinct tails were observed. In some cases, the maximal registered inter-arrival time was three orders of magnitude larger than the supposed value. The use of better hardware and newer software versions helps to improve the accuracy but, still, it does not reach the precision of DAG. Thus, despite the use of high-end machines for running tools like `tcpdump` and `windump`, there remains the need to check whether the quality of the time stamping matches the needs of the subsequent analysis.

3.3 Traffic Characterization

The following subsections contain a broad realm of qualitative studies of traffic behaviors in different scenarios. Multicast, Optical Burst Switching (OBS), and TCP traffic are investigated with regard to self-similar properties. Further, traffic aggregation for modem pool traffic yielding Gaussian properties is discussed. Finally, General Packet Radio Service (GPRS), Wireless LAN (WLAN), and Peer-to-Peer (P2P) traffic are characterized.

3.3.1 Multicast Traffic

Most efforts on multicast communication have focused on developing new applications and protocols that are compared with respect to performance measures such as scalability, reliability, and congestion control. However, the nature of the traffic stream generated by each type of protocol, particularly with respect to self-similarity, has not been studied extensively. The transport layer mechanisms are important components in translating heavy-tailed file size distributions at the application layer into link level traffic self-similarity. In particular, [93] and [94] focus on the traffic that scalable multicast protocols generate. The results provide empirical evidence for the fact that the level of self-similarity depends on the transport protocol used. Bimodal Multicast is a novel option in the spectrum of multicast protocols. It is based on an epidemic loss recovery mechanism which has a Markovian structure. The protocol has been shown to exhibit stable throughput under failure scenarios that are common on real large-scale networks. In contrast, this kind of behavior can cause other reliable multicast protocols to exhibit unstable throughput. Bimodal Multicast is compared to Scalable Reliable Multicast (SRM), as the latter is inspired by the principles of the TCP/IP protocol stack, which is prevalent in the Internet and exhibits self-similarity.

In [93], empirical results demonstrate that the epidemic approach of Bimodal Multicast generates more desirable traffic than SRM in the case of a Constant Bit Rate (CBR) source. SRM traffic shows long-range dependence and self-similarity, whereas Bimodal Multicast traffic is short-range dependent. The approach is to analyze simulation traces obtained from the `ns-2` simulator and provide analytical support for the results. The delays at the transport level and traffic at the link level are studied. Bimodal Multicast yields lower overhead traffic and transport delays than SRM. The protocol mechanisms are elaborated on as the underlying factor in the empirical results. The intrinsic relation of these mechanisms to traffic characteristics is studied through delay analysis in the case of Bimodal Multicast and discussed for SRM. Figure 3.6 shows the striking difference between the values of the Hurst parameter obtained from the delays generated by the SRM and Bimodal Multicast protocols for CBR and ON/OFF traffic as the group size scales up.

The marginal delay distribution has been analyzed for Bimodal Multicast and is shown to decay exponentially. This substantiates that Long Range Dependence (LRD) is not expected intrinsically, due to the epidemic mechanism of the protocol. More information on epidemic loss recovery can be found in Section 1.9.

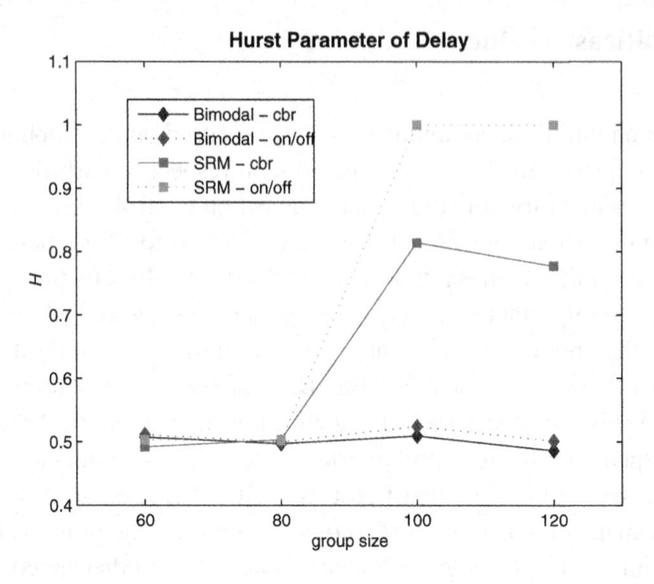

Figure 3.6: Comparison of Hurst parameters of the delay for Bimodal Multicast and SRM as a function of the group size

In [94] is considered a self-similar source, namely an ON/OFF sender that transmits with Pareto ON and OFF times. It is well known that when sufficiently many traffic streams from such sources are aggregated, long-range dependence arises at the link level. Results pertaining to self-similar sources are compared with the CBR case. In the case of an ON/OFF sender, Bimodal Multicast still generates short-range dependent delay, which is scalable in the number of users, thus showing superior performance properties in this respect, in addtion to its other superior performance properties of stable throughput, lower protocol overhead, and higher reliability. However, long-range dependence emerges at the link level when the source induces self-similarity. Although there is a single ON/OFF source, at the link level there is an aggregation arising from the recovery process of all receivers in the network. For SRM, the traffic becomes worse in terms of both delays and at the link level compared to the CBR case. The timer-based loss recovery mechanism of SRM triggers self-similarity with long-range dependence, whereas the Markovian structure of Bimodal Multicast inherently generates short-range dependent traffic. As future work, Bimodal Multicast deserves more research in terms of traffic characteristics for various parameters, such as gossip rate and buffer sizes. One goal for that work is to provide a stochastic model involving the parameters and the mechanisms of Bimodal multicast. Comparative studies with other scalable multicast approaches will help to identify efficient protocol mechanisms.

3.3.2 Optical Burst Switching Traffic

In Optical Burst Switching (OBS) networks, the essential function of *burst assembly* puts together, at the edge nodes, a number of IP packets into an optical burst of larger size so as to achieve efficient burst switching. The ability of OBS to reduce the self-similarity of IP traffic, heavily discussed recently, is important for the evaluation of OBS network performance. In [199] this question is studied both via simulation and via theoretical analysis. Three different assembly schemes are inspected: threshold-based scheme, time-based schemes, and time-based schemes with padding. The traffic is studied in terms of both byte stream and packet, or burst, stream. It is found that only in the case of time-based assembly and regarding the burst departure process, self-similarity is reduced when increasing the time-out interval of the assembly algorithm. In all other cases, especially when traffic is measured as byte-stream, assembly has no impact on the self-similarity properties.

3.3.3 TCP Traffic

A measurement and analysis study is presented in [186] and [200] on how TCP congestion control can propagate self-similarity between distant areas of the Internet. This property of TCP is due to its congestion control algorithm, which adapts to self-similar fluctuations on several timescales. The mechanisms and limitations of this propagation are investigated, and it is demonstrated that if a TCP connection shares a bottleneck link with a self-similar background traffic flow, it propagates the correlation structure of the background traffic flow above a characteristic timescale. The cut-off timescale depends on the end-to-end path properties, e.g., round-trip time and average window size. It is also demonstrated that even short TCP connections can propagate long-range correlations effectively. The analysis reveals that if congestion periods in a connection's hops are long-range dependent, then the end-user perceived end-to-end traffic is also long-range dependent and is characterized by the largest Hurst exponent. Furthermore, it is shown that self-similarity of one TCP stream can be passed on to other TCP streams multiplexed together with it. These mechanisms complement the widespread scaling phenomena reported in a number of recent papers. The arguments are supported with a combination of analytic techniques, simulations, and statistical analyses of real Internet traffic measurements.

In [192] are investigated statistical properties of a TCP flow-interarrival process using the tools `Tstat` and `DiaNa`. To that end, the original trace is split into subtraces, each of which has the same amount of traffic in Bytes and belongs to the same origin/destination pair. Furthermore, a partition into large relations ("elephants") and small relations ("mice") is performed.

3.3.4 Modem Pool Traffic

Gaussian models are discussed in [201] and in its preliminary version [185]. Whereas [185] is more oriented to traffic characterization, [201] turns more to the methodological direction. Any Gaussian traffic model is motivated by the Central Limit Theorems (CLTs). Hence, to justify Gaussian models there must be enough traffic aggregation. The aggregation can be theoretically divided into the *horizontal* and *vertical* directions. Horizontal aggregation is the time scale or resolution, i.e., the width of time slots, and vertical aggregation is essentially the number of contributing sources at a time slot.

The traffic trace analyzed in [185] and in [201] was measured from a modem pool of a commercial Internet Service Provider (ISP). Since the level of vertical aggregation was rather high, a natural question to ask was whether it was high enough to justify Gaussian traffic models for some, or any, time resolutions. In [185] are contained already the basic ideas of determining the minimum time scale and minimum number of contributing sources that would be required by the CLT. In [201] some simple and pragmatic ideas were used to determine the first time scale where the Gaussian approximation is plausible. In the downstream directed traffic this was between 0.512 s and 1.024 s for the data trace studied. However, the effect of the CLT was already visible from a 0.128 s resolution, but the amount of vertical aggregation was not sufficient for resolutions from 0.128 s to 0.512 s. Some heuristic approximations were made to determine how much more vertical aggregation would be required in these smaller time scales in an ideal case. In the upstream directed traffic none of the timescales studied gave satisfactory results, and some reasons for this are also explained in [201].

3.3.5 GPRS Traffic

GPRS traffic is considered in [202] and in its preliminary version [188]. These papers present some results obtained from `tcpdump` traces recorded between the GPRS backbone network and the Internet, but before the Network/Private Address Translation (NAT) is done. The measurement was done in May, 2002, about half an year after the operator had launched its commercial GPRS service. It is probably one of the first published GPRS user traffic measurements in the world. A GPRS session was defined as all the packets, and associated time stamps, with the same temporary IP address that the user is given when he/she is attached to the GPRS network. The main observations were that during the measurement time, most GPRS sessions started during the working days and in working hours, that they were in general roughly similar to low access speed dial-up sessions, and that session durations and session volumes seem to have heavy-tailed distributions. Also, the majority of the data transfer occurs typically in the beginning of the session and the GPRS users typically detach from the GPRS network when they have finished active usage.

In [189] are studied distributional properties of GPRS/GSM session volumes and durations, by combining a known statistical method of ascertaining about the underlying distribution with the data-analysis of GPRS data introduced in [188] and in [202]. The statistical method is based on the concept of maximum correlation, and was used because there was no *a priori* reason to expect or prefer any particular distribution. The main result is that these data sets exhibit a heavy-tailed nature. The Weibull distribution with the shape parameter between $0 < \alpha < 2$ is one of the possible models. A first draft version of the document is contained in [189].

3.3.6 WLAN Traffic

In [182], a wireless next generation IP-network and its corresponding QoS management mechanisms are considered. A series of comprehensive traffic measurements is used to study the interworking of flow control, resource reservation, and mobility management. Considering a terminal roaming in a basic service area of a 2 Mbit/s IEEE802.11 WLAN, typical results of the RTT as compared to the SNR and related handovers, and the evolution of the TCP congestion window, are shown in Figure 3.7 and 3.8, respectively.

Two recent studies [195, 177] investigate the impact of WLAN on transport level, i.e., TCP and UDP, traffic performance. They present time plots revealing the interaction between WLAN link layer parameters, such as SNR as a function of the localities or the preset bit rate, and TCP parameters, such as perceived throughput, RTT, server window, and retransmissions. Details are reported in Section 4.4.2.

3.3.7 Peer-to-Peer Traffic

Although P2P is currently mainly related to doubtful “royalty free” access to resources, P2P file sharing still generates a significant amount of traffic in networks [203]. A key characteristic of P2P applications is the strong fluctuation of P2P traffic patterns in time and space. As a result, it is anticipated that traditional network design techniques and traffic engineering procedures may no longer be applicable and new methods may be needed that maintain the autonomous and self-organizing characteristics of P2P.

In [197] is presented a measurement study on the signalling traffic in Gnutella overlay networks. Both signalling load and the scale of variability in the existence of P2P overlay connections are investigated. The signalling traffic in

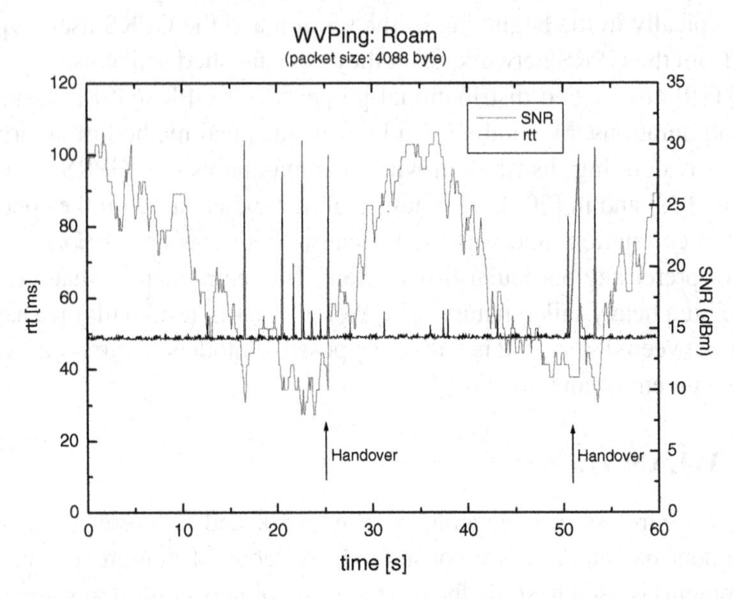

Figure 3.7: RTT delay samples from `WVPing` (cf. Section 3.2.1) with 4088 byte payload for roaming in a basic service area of the 2 Mbps environment

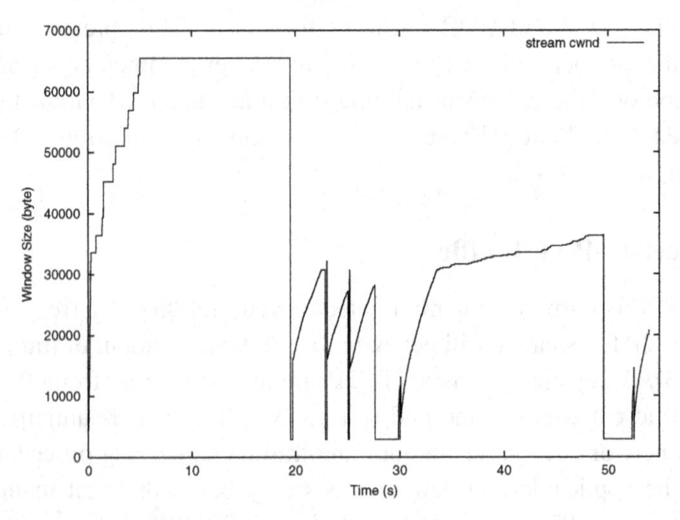

Figure 3.8: Dynamics of the TCP congestion window subject to roaming in one basic service area for stream traffic patterns in the 2 Mbps environment

Gnutella overlays varies significantly over short timescales, due to the Gnutella use of flooding protocols. The investigation of the overlay connection holding time in Gnutella showed that the distribution typically has bi-modal characteristic. The modes identify the time scales on which a dynamic and adaptive management of P2P overlays and P2P services is of advantage or needed.

In [191] is provided a traffic profile for the eDonkey P2P filesharing service [196]. The traffic profile shows that signalling and download have significantly different characteristics. A future traffic model has to distinguish between the two types of traffic. In addition, the traffic profile, provided in [191], gives evidence that the expected "mice and elephant" phenomenon in eDonkey traffic is not as severe as expected.

A more detailed discussion on these findings is found in Section 6.3.

3.4 Traffic Modeling

This section starts with contributions on the estimation of probability density functions (pdf), which is an important initial step in the process of modeling. Several applications of Poisson- and Markov-type models and their matching to real data are discussed. Among others, long-range dependent and transient behavior is considered. Following, time-discrete models serving as basis for many queueing studies within COST 279, c.f. Section 2.2, are introduced. Furthermore, flow-level models, their parameters, and applicability are discussed. The subsection on fluid models deals with the impact of modeling timescales on performance, and the capability of detecting and describing the impact of bottlenecks. Thereafter, a couple of fractal-type models is presented. Finally, hints to mobility models are given.

3.4.1 PDF Estimation

In [204] is studied the estimation of heavy-tailed probability density functions, their mixtures, and high quantiles. First, the relevance of this issue in teletraffic engineering is discussed and then a new combined estimation technique for such pdfs f is proposed. The "tail" of the pdf is estimated by a parametric tail model $f_\gamma^T(x) = \gamma x^{-\gamma-1} + 2\gamma x^{-2\gamma-1}$, and its "body" $f^B(t) = \frac{1}{x_{(k)}} \sum_{j=1}^{N} \lambda_j \varphi_j(\frac{t}{x_{(k)}})$ by a non-parametric method in terms of a finite linear combination of trigonometric functions $\varphi_k(t) = \sqrt{\frac{4}{\pi}} \cos\left((2k-1)\frac{\pi}{2}t\right)$ that are evaluated at the data points $x_{(k)}$ of a sample. In order to minimize the mean-squared estimation error, the parameters of the parametric and non-parametric parts are estimated by means of the bootstrap method and the structural risk minimization method. The latter parameters are determined by the number of extreme-valued data that are used in Hill's estimate of the tail index γ and the number N of terms and coefficients of the linear combination. The new method is illustrated using some relevant mixtures of heavy-tailed pdfs, e.g., a mixture of a Gamma and a Pareto distribution arising from the delay modeling of IP traffic, and applied to construct a high quantile estimate. Furthermore, its effectiveness is shown by an application to real data derived from Web-traffic characteristics like session durations and volumes.

Measurements of Web traffic have shown that its characteristics like session durations, transferred volumes, and file sizes are governed by heavy-tailed distributions $F(x)$. In [205], the new task of Web data mining to estimate the underlying heavy-tailed pdfs $f(x) = F'(x)$ by on-line algorithms is addressed. To guarantee a better estimation of the tail behavior, a new specific transformation scheme $T_{1/\gamma} : X \rightarrow [0, 1]$ is proposed. It is adapted to the empirical data $X = \{X_1, \ldots X_n\}$. Applying the latter, a new on-line estimator for such pdfs is developed as a further basic innovation. In this scheme the required extreme-value index $1/\gamma$ of the pdf f is reconstructed by a new recursive technique.

In [47] are tested empirical flow statistics against commonly used theoretical probability distributions, such as hyper-exponential, Gamma, Weibull, Pareto, and log-normal. It is also discussed the suitability of statistical tests, such as the Kolmororov-Smirnov test and the Chi-Square test, for these kinds of tasks.

3.4.2 Poisson- and Markov-Type Models

Poisson Cluster and Transient Markovian Arrival Processes for Web Traffic

In [206] is studied a model of HyperText Transfer Protocol (HTTP) sessions initiated at a Web server. The model is based on a previous analysis carried out by Liu et al. [207], among others, where one HTTP session is composed of one main object and possibly several in-line objects downloaded in parallel. The HTTP session finishes when the last downloading process ends. The objects of the study are the stationary distribution of the total number of objects which are currently downloaded and its covariance function. The model is a particular case of a Poisson cluster process and is more precisely defined as follows. At the Web server, HTTP requests occur according to a Poisson process. To each HTTP request is associated a process of object requests determined by a *transient Markovian Arrival Process* (tMAP), introduced in [208]. The Markovian assumption has the advantage of making the whole process very tractable and allows for much flexibility in modeling phenomena such as high correlations over long intervals of time [209], even though Markovian models do not exhibit long-range dependence stricto sensu. To each object is associated a mark, namely its downloading duration. We assume that the durations are independent and identically distributed (iid) random variables with a common distribution; this implies that the capacity of the system is sufficiently large that a transmission is not slowed down by other downloading processes which might be in progress at the same time.

A procedure for model fitting is then constructed in [178]. This procedure determines the input parameters of the tMAP according to observed arrival times of object requests within an HTTP session. It is build on the so-called *Expectation Maximization* (EM) *algorithm*. The name EM algorithm stems from the alternating application of an expectation step and a maximization step which yield successively higher likelihoods of the estimated parameters.

Equivalent Poisson Traffic

In the framework of inter-satellite routing, [210] presents an empirical traffic source model derived from an Internet backbone traffic trace, which is then compared to equivalent Poisson traffic as a point of reference. The comparison is done using traffic class dependent routing, which has the potential to differentiate between traffic classes using different optimization criteria in route calculation. Interestingly enough, the performance measures based on aggregate traffic flow show no significant difference between routing of empirical and equivalent Poisson traffic.

Markov Modulated Poisson Processes

Markov Modulated Poisson Processes (MMPPs) are very popular traffic models due to their ease of use, mainly due to their additive property, and to the availability of analytical results for the evaluation of their queueing behavior. In [211] is proposed a parameter fitting procedure using MMPPs that leads to accurate estimates of queueing behavior for network traffic exhibiting LRD behavior. The procedure matches both the autocovariance and marginal distribution of the counting process. A major feature is that the number of states is not fixed a priori and can be adapted to the particular trace being modeled. The MMPP is constructed as a superposition of L 2-MMPPs and one M-MMPP. The 2-MMPPs are designed to match the autocovariance and the M-MMPP to match the marginal distribution. Each 2-MMPP models a specific time-scale of the data. The procedure starts by approximating the autocovariance by a weighed sum of exponential functions that model the autocovariance of the 2-MMPPs. The autocovariance tail can be adjusted to capture the long-range dependence characteristics of the traffic, up to the time-scales of interest to the system under study. The procedure then fits the M-MMPP parameters in order to match the marginal distribution, within the constraints imposed by the autocovariance matching. The number of states is also determined as part of this step. The final MMPP with $M \cdot 2^L$ states is obtained by superposing the L 2-MMPPs and the M-MMPP. The inference procedure was applied to traffic traces exhibiting long-range dependence, and its queueing behavior was evaluated through simulation. Very good results were obtained for the traces used, which include the well-known Bellcore traces, both in terms of queueing behavior and number of states.

In [190] is proposed a simple MMPP traffic model that approximates the LRD characteristics of traffic traces measured at an edge router, at both flow and packet level. The MMPP model mimics the real behavior behind the interaction between users, protocols, and the network, using the notion of sessions and flows, therefore resulting in a simple and intuitive model. The estimated Hurst parameter of the synthetic traffic fits the one observed from data traces measured both at flow and packet level. The characteristics of the synthetic traffic generated with the model match the LRD characteristics observed in the measured traces over the time scales of interest. One of the interesting features of the proposed MMPP model is that it requires only five parameters. Three of these parameters can be directly mapped onto average traffic parameters, such as the average flow arrival rate, the average number of packets per flow, and the average arrival rate of packets within flows. The other two parameters define the notion of session, and are used to control the Hurst parameter of the synthetic traffic on the considered scaling range. The queueing behavior, for finite and infinite buffer queues, of the traffic generated by the model is coherent with the one of the measured traces at several different traffic loads. While the model is not intended to offer an explanation of the reasons why Internet traffic is LRD, it does offer a simple and manageable tool for dimensioning and planning e.g., link and buffer capacities in networks, since the characteristics of the generated traffic are easily controlled through the model input parameters. The simplicity of the MMPP traffic model makes it an ideal traffic generator to drive simulations and, in addition, its Markovian properties makes it analytically tractable, so that analytic solutions for simplified networking scenarios are possible.

3.4.3 Time-Discrete Models

In general, time-discrete models describe the amount of workload usually assumed to be iid and arriving or processed during one time slot of arbitratry size. Such models have served for many queueing studies within the COST

Action 279, cf. Section 2.2.

Here, the use of these models is exemplified by a specific contribution related to traffic characterization. In [165], the evolution of the interarrival and interdeparture times between voice packets when they are proceeding through a number of network nodes has been investigated. Instead of separately specifying the characteristics of each individual source, the arrival process in a node is modeled as the superposition of a single tagged stream and an independent background process that aggregates the remaining traffic sources. Because the load of a single voice stream is assumed to be very low compared to the load of the aggregate traffic, the tagged voice packets can be represented as markers, i.e., packets with size zero. The tagged marker stream is characterized by the consecutive interarrival times between the markers, which are assumed to be identically distributed. The background arrival process on each network node is described on a slot-per-slot basis according to a general iid process. The results indicate that the mean interarrival time between markers remains constant throughout the network, and that both the variance of the interarrival times and the covariance between two successive interarrival times are increasing functions of the number of stages traversed by the markers. Also, it has been observed that if at the entrance of the first network stage two successive interarrival times are independent of each other, after one stage they become negatively correlated.

3.4.4 Flow-Level Models

Given the complexity of Internet traffic at packet level, some performance models are currently developed, particularly for elastic traffic under TCP control (see [61] and [59]), that consider higher level entities, namely the traffic "flows" or "sessions." In order to investigate several assumptions commonly considered in such models, a detailed statistical analysis of the IP traffic processes at flow level is performed in [47], based on traces gathered on a wide area network. First, the concept of flow is thoroughly discussed and a method is suggested to estimate an adequate Time Out value to distinguish different flows within a traffic stream. Next, the main result of this contribution is the rejection of the Poisson hypothesis for the flow arrival process: it is shown that Gamma and Weibull distributions, the latter in agreement with [212], provide excellent fits to the empirical flow inter-arrival time distributions; in addition, as can be seen on Figure 3.9, some consistent level of correlation is detected between successive flow arrivals. Finally, the study confirms the well-known heavy-tailed character, mostly Pareto-like, of flow size distributions, both for UDP and TCP traffic.

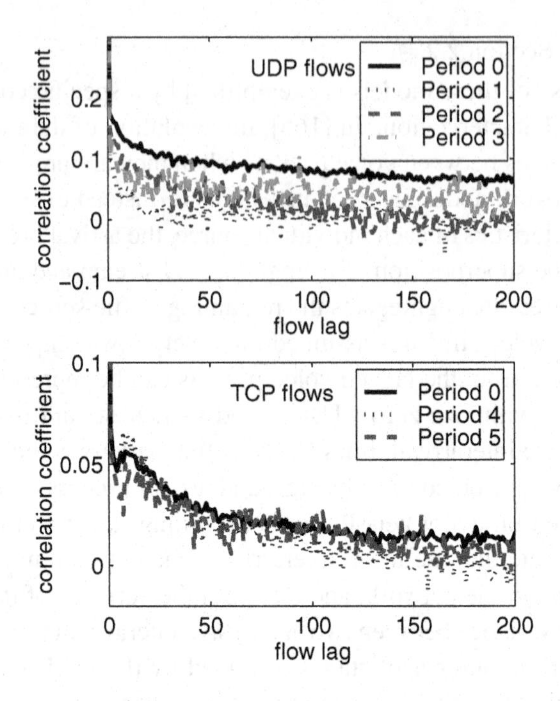

Figure 3.9: Autocorrelation functions of UDP (upper) and TCP (lower) flow inter-arrival time

To evaluate the performance in scenarios of integration of streaming (voice, video) and elastic (data) traffic in a multiservice network, an analytical model is presented in [62]. Assuming Poisson flow arrivals on a network link and fair bandwidth sharing for elastic flows, the model is based on the performance of an M/G/1 Processor Sharing (PS) queue with time-varying capacity. The variable capacity is that which is left available by streaming flows, the packets of which are supposed to have priority with respect to those of elastic flows. More details are found in Section 1.5.4.

A related scenario is considered in [61] and [59], where two classes of elastic TCP traffic are distinguished, i.e., premium (high priority) and best effort; the TCP flows have a limited peak rate r. This scenario is modeled by a multiple server PS queue with with two priority classes. The analysis is presented in [61]. In particular, for the high priority traffic the mean sojourn time (i.e., file transfer time) is obtained from known analytical results for the M/G/c PS queue in [48]. For the best effort traffic, a simple accurate approximation for the mean sojourn time is derived. In [59], the analytical modeling results of [61] are compared with results obtained by `ns` simulations of the scenario described above.

With regards to TCP performance, PS models capture quite "roughly" the statistical multiplexing effect on flow level and allow for link dimensioning. However, these models do usually not provide insight into the impact of packet level parameters such as the round trip time and buffer size. Therefore, in [213] an *integrated* packet/flow level model for the analysis of a bottleneck link in a TCP/IP network environment is developed, combining the attractive features of both flow level and packet level models. The approach works roughly as follows. First, for a given number of flows, say n in the system, the throughput t_n is computed. This is done by using well known packet level TCP models reflecting the impact of round trip time and buffer size (see, e.g., [214]). Next, on the flow level, the transfer times are obtained from a processor sharing model with Poisson flow arrivals and state dependent service rates t_n [48]. An interesting feature of this model is that the mean flow transfer time depends on the flow size distribution only through its mean value. This *insensitivity* property is confirmed by simulation.

In [144] is presented a similar integrated, hybrid packet/flow level modelling approach, but for the analysis of flow throughputs and transfer times in IEEE 802.11 WLANs. While the packet level model covers the statistics of the packet transfer at MAC level, the flow model reflects the system dynamics in terms of initiation and completion of flows. The resulting integrated model is analytically tractable and yields accurate approximations for throughput and flow transfer time values. In particular, a similar insensitivity property for the mean flow transfer time is obtained as in the TCP model of [213] discussed above.

3.4.5 Fluid Models

The impact of approximating packetized flows by fluid flows on the Complementary Cumulative Distributions (CCDs) of the unfinished work is investigated in [121]. For that purpose two discrete-time queueing models are compared. In the first model, referred to as the packet model, two timescales are present in the input traffic, i.e., a burst and a packet timescale. In the second one, referred to as the fluid model, only an (identical) burst timescale is present. It is assumed that time is divided in units of constant length, called frame times, and that every frame time is further divided into constant length time units, called packet times. As a typical model for a bursty traffic source, a two-state discrete-time Markov source is considered. This Markov source has a frame time as underlying time unit, i.e., the source can only change state on frame time boundaries. During a frame time in which the source is in a given state, it generates a certain amount of bytes. In the packet model, these bytes are divided into fixed-length packets which are sent bunched up in the beginning of the frame time. In the fluid model the bytes are sent at a constant rate over the whole duration of the frame time. A queueing model with an infinite buffer is then considered in which the traffic of either M identical packet sources or M identical fluid sources with the same parameters is multiplexed. The performance measure of interest is the distribution of the amount of unfinished work in both systems. Numerical results show that the slope of the CCDs of the unfinished work obtained with both models is the same, i.e., fluctuations in the arrival pattern at the smallest timescale do not influence this slope. The probability that the amount of unfinished work is larger than a certain value is the smallest for the fluid model, which neglects traffic fluctuations at the finest timescale. The difference between the two CCDs becomes however less important when the parameters of a scenario are changed such that the slope of the curves increases, i.e., when the probability of having a larger amount of unfinished work in the system becomes larger. Several properties of traffic scenarios that result in a larger slope of the curves are considered in [121]. In [215], the model presented in [121] is used in a case study dealing with a video streaming application.

The impact of time scales on the link capacity required for maintaining a desired QoS is discussed in [57], cf. Section 1.5.3.

In [131], the stochastic fluid flow model demonstrates the capability to reveal the impact of bottlenecks, i.e., shortages in capacity, on packet streams. This is achieved by comparing throughput histograms at the output with those at the input of the bottleneck. Assuming an Anick-Mitra-Sondhi (AMS)-type fluid flow model [125] and that the joint probability of an (non-)empty buffer in each state of the model can been calculated, [131] shows how to obtain the output bit rate distribution both for individual traffic streams and in total. From these comparisons, it can be seen whether there is other interfering traffic sharing the bottleneck or whether the bottleneck has a buffer of significant size. This allows for both identification and classification of the bottleneck. Under certain conditions, the capacity of the bottleneck is revealed even in the throughput histogram of an individual stream at the output of the bottleneck. For the particular example presented, it is observed that streams with different characteristics (constant or variable rate) inherit the same kinds of changes in their throughput histograms. An application of this model for bottleneck classification [183] is presented in Subsection 3.5.3.

A fluid flow source model displaying multi-fractal properties is used in [127] and will be presented in Section 3.4.6.

3.4.6 Fractal-Type Models

It has been observed in several measurements and statistical studies that network traffic exhibits fractal properties. Deep analyses have shown that network traffic can also present multifractal characteristics. However, so far only a few relevant multifractal models have been developed and these are based on constructing a multiplicative process structure.

A fluid flow source model displaying multi-fractal properties was proposed in [216] and is used in [127] to model data traffic. These properties are obtained by multiplying the outputs of a number N of Markov Modulated Rate Processes (MMRPs), each of which is acting on a separate timescale. Actually, one might think of one "fast" MMRP whose rate is modulated by "slower" MMRPs incorporating activities on different time scales. As N is finite, the whole process that is showing long-range dependence on short time scales becomes short-range dependent on long time scales. In [127] are presented time plots and variance-time log-log plots for different parameter settings.

Multifractal behavior was recently observed in several traces of IP Wide Area Network (WAN) traffic. In [217] is proposed a novel traffic model that is able to capture multifractal behavior and characterizes jointly the processes of packet arrivals and packet sizes. The construction of the traffic process is based on stochastic L-Systems, introduced by the biologist A. Lindenmayer as a method to model plant growth. The model is characterized by an alphabet, an axiom, and a set of production rules. The alphabet is a set of symbols. The production rules define transformations of symbols into strings of symbols. Starting from an initial string, the axiom, an L-System iteratively constructs sequences of symbols, replacing each symbol by the corresponding string according to the production rules. The traffic model uses a single L-System alphabet and production rule, where the alphabet is a set of pairs, and each pair element represents a packet arrival rate and a packet mean size. In this way, the model is able to capture correlations between arrivals and sizes, leading to an accurate prediction of the queueing behavior. A detailed comparison with a related multifractal model based on conservative cascades is provided in [217]. The results, which include applying the fitting procedure to real observed data with multifractal scaling behavior on both the packet arrival and packet size processes, show that the L-System based model can achieve excellent fitting performance in terms of first and second order statistics and queueing behavior.

In [187] is presented the new monofractal model *Limit of the Integrated Superposition of Diffusion processes with Linear differential Generator* (LIS-DLG), which is not self-similar, and has a more powerful modeling ability to capture scaling behavior when compared to self-similar models. It is argued that the monofractal model is flexible enough to accurately capture the fractal scaling of network traffic, and that there is no need to use more complex multi-fractal models. In this model the cumulants of the measured traffic are fitted to the cumulants of the process generated by the model, and it is shown that the resulting bispectrum of the traffic can be accurately captured. Several properties of the LISDLG model are presented in the report, including covariance structure, cumulants, spectrum, and bispectrum. The relevance and validation of the proposed model are demonstrated by application studies for measured Internet traffic.

3.4.7 Mobility Models

In mobile and ad-hoc scenarios, mobility models capture the location of moving nodes sending and receiving traffic. They thus constitute an important base for assessing the performance of mobile and ad-hoc networks, e.g., in terms of connectivity and/or hot spot behavior. In [218] and [219] is considered the *Random Waypoint Mobility Model*, which starts with a random or given initial distribution of the nodes in space, and where nodes subsequently move in straight lines from each waypoint to a new randomly chosen waypoint, at an also independently randomly chosen uniform speed.

3.5 Interpretation and Application

This final section focuses on the challenges of interpreting measurements and using the results of measurements for operational tasks. A framework for the interpretation of measurement and examples from the area of traffic estimation and classification are presented. Moreover, applications of measurement results for traffic prediction and traffic control are described.

3.5.1 An Interpretation Framework

Motivated by the fact that a large number of published papers in empirical networking analysis follow a generic framework that might be formalized and generalized to a large class of problems, [176] introduces a framework for interpreting measurement obtained over Internet. It aims at providing researchers being confronted with measurements coming from a network with some guidelines on how to formalize the way to address interpretation of observations.

Interpretation is essentially a matter of relating observed effects to hidden causes. This problem might be formalized in its most general setting as an inverse statistical inference problem. In a first step, the constructive approach (cf. Section 3.1) is used to define some models relating effects to causes, or differently said, relating parameters of unseen, i.e., hidden input scenarios to observed and measured parameters. This constructive model will be used as an a-priori base for interpretation when the descriptive approach and statistical inference are used to infer the hidden input scenario that have led to the actual observations. The inverse statistical inference problem can be solved by applying the EM method or the Bayesian framework.

This interpretation framework is illustrated by well-known examples from the networking literature, namely network tomography (cf. Section 3.5.2) and interpretation of active measurements (cf. Section 3.5.3). It demonstrates that even if at first glance these two problems are different, the solution framework is the same.

3.5.2 Traffic Matrix Inference

A Traffic Matrix (TM) reflects the volume of traffic that flows between source and destination nodes in a network. The nodes can refer to a variety of network elements such as POPs, routers, or even address prefixes. A POP-to-POP traffic matrix X captures the amount of traffic exchanged between two POPs, where X_{ij} represents the volume of traffic traveling from ingress POP i to egress POP j. The value of X_{ij} usually represents a bandwidth value averaged over some time interval, although other types of elements are also possible. There are a number of traffic engineering tasks that could be greatly improved with the knowledge provided by traffic matrices. Capacity planning, routing protocol configuration, definition of load balancing policies, and failover strategies are tasks that would benefit from having information on the size and locality of traffic exchanges. An important example is the setting of OSPF or IS-IS routing weights. With knowledge of the TM, an algorithm for setting weights will select a routing that achieves a significantly better load balancing than one with an incorrect idea of the TM.

Obtaining a traffic matrix can be basically approached in two ways. One may directly measure it, or one can rely on partial information to infer it. Measurement approaches have not been fully explored because they involve overcoming challenging engineering obstacles related to the deployment of a measurement infrastructure, and to the storage and processing of large amounts of information. Furthermore, the monetary cost may be high. Instead, previous work on obtaining traffic matrices has relied on statistical inference techniques that use partial information to estimate the TM. The term Network Tomography has been coined for this problem when the partial data come from repeated measurements of the traffic flowing along directed links in the network. Such data are usually obtained from SNMP, which allows measuring the total amount of incoming and outgoing bytes on a link, typically over five-minute intervals. The idea behind inference approaches is to use these link statistics to infer the characteristics of end-to-end flows. End-to-end flows are defined within a single domain and are usually referred to as Origin-Destination (OD) pairs. In a POP-to-POP topology, the origin and destination nodes are POPs.

In addition to inference methods, it is also possible to formulate the traffic matrix estimation problem as a constrained optimization problem and use techniques such as Linear Programming (LP). In [220] is presented a comparative study of existing TM inference techniques. The evaluated statistical techniques are found to outperform an LP-based technique, still statistical techniques are significantly restricted in their ability to converge to the right solution. This is because they rely on scarce actual network information and they require intensive computation to reach reasonably accurate estimates. These restrictions impose a substantial burden on the quality of the starting point that should be provided to guide the estimation process. In [220] is used a very fast variant of the EM algorithm for the network tomography problem. The improvements made are aimed at reducing the computation requirements of the algorithm, enabling it to expand the iterative horizon in search of global optima as solutions to the inference problem. Second, alternative modeling approaches are investigated to provide reasonable starting points for inference techniques. These starting points are called informed priors because they are obtained from models that incorporate substantial network information. The paper introduces choice models for generating informed priors. Different approaches are compared with respect to the estimation errors yielded, sensitivity to prior information required, and sensitivity to the statistical assumptions they make. The impact of characteristics such as path length and the amount of link sharing on the estimation errors is also studied.

3.5.3 Estimation and Classification

The above mentioned methodology is applied in [180] to cross traffic estimation for analyzing and interpreting measurement collected over the Internet. As pointed out in Section 3.5.1, the approach is based on inferring cross traffic characteristics that lead to the observed losses by using an associated a priori constructive model. The constructive model used in [180] is an MMPP/M/1/N single server bottleneck. The originality of this solution is that it starts with the observed loss process and infers inputs that have led to these observations. The methods presented in [180] provide a powerful solution to address the complexity of interpreting IP active measurement and empirical network modeling.

In [183] is presented a study aiming at classifying bottlenecks with the aid of throughput histograms. The study was carried out on UDP packet streams carrying voice and video between Karlskrona, Sweden, and Würzburg, Germany. These videoconferencing streams passed a bottleneck, a 10 Mbps half-duplex link that also carried a disturbing UDP stream of varying intensity. As bottleneck indicator, throughput histogram difference plots were proposed. These throughput histograms stem from bit rate measurements over averaging intervals ΔT during an observation interval ΔW, at a throughput resolution of ΔR. The constructive model used as basis for the interpretation of the results was a stochastic fluid flow model as presented in [131]. Based on this model, the classification of the impact of the bottleneck caused by the cross traffic on the individual voice and video streams was possible. Depending on the intensity of the cross traffic and the actual interaction with the voice and video streams, the latter realized either a *shaping bottleneck*, i.e., one where the variations in throughput were damped by the bottleneck, or a *shared bottleneck*, i.e., one where the variations of the throughput grew due to interaction with the interfering traffic, exactly as postulated by [131].

In order to control and manage highly aggregated Internet traffic flows efficiently, we need to be able to categorize flows into distinct classes and to be knowledgeable about the different behavior of flows belonging to these classes. In [221] is considered the problem of classifying Border Gateway Protocol (BGP) level prefix flows into a small set of homogeneous classes. Using the entire distributional properties of flows can have significant benefits in terms of quality in the derived classification. A new method based on modeling flow histograms using Dirichlet Mixture Processes for random distributions is presented, together with an inference procedure based on the Simulated Annealing Expectation Maximization algorithm that estimates all the model parameters as well as flow membership probabilities, i.e., the probabilities of flows belonging to classes. The method is capable of examining macroscopic flows while simultaneously making fine distinctions between different traffic classes. Furthermore, it can cope with flows being close to class boundaries and the inherent dynamic behavior of Internet flows.

Yet another important task, e.g., regarding application areas like security and traffic Engineering, is application recognition. Well-known port numbers can no longer be used to reliably identify network applications. There is a variety of new Internet applications that either do not use well-known port numbers or use other protocols, such as HTTP, as wrappers in order to go through firewalls without being blocked. One consequence of this is that a simple inspection of the port numbers used by flows may lead to the inaccurate classification of network traffic. Moreover, because of privacy concerns or more simply because of used encryption mechanism, it is frequently impossible to get access to the full payload of packets. This means that classification should be based only on the behavior of the packet flow in term of size, interarrival time and interaction. In this context, [222] presents a blind applicative flow recognition through behavioral classification. The approach is based on very simple sequences of quantified packet size and packet direction. These sequences are clustered through a powerful spectral clustering algorithm. Thereafter is developed a recognition algorithm based on a mixture of Hidden Markov Models (HMM) representative of the obtained clusters. The presented method appears to be very powerful, as it reaches a recognition performance of 90 % with only observing seven packets of a flow. This work is a first step toward an operational flow recognition system robust toward flow morphing, i.e., tunneling a flow in another protocol and payload encryption.

3.5.4 Traffic Prediction

In [223] is considered the usability of the traditional linear predictors, especially in the case where the number of aggregated flows is moderate or small. First are stated some theoretical results based on the fractional Brownian motion, then real traffic traces with different aggregation levels are analyzed. Finally, some comments on the prediction based dynamical resource reservation are given. The main results can be summarized as follows. If the traffic has a power law-type variance structure and possibly non-stationary features, then the combined use of a moving average mean rate estimate and a fixed Fractional Brownian Motion (FBM) type variance function is motivated. This results in a robust and pretty good linear predictor. In the case of a prediction delay, a working engineering solution is to condition with respect to four or five geometrically increasing intervals with smallest interval about half the prediction interval. However, there are situations when the simple mean rate estimation is enough, i.e., it cannot be improved by more complicated algorithms. Moreover, a straightforward application of traffic predictors in resource reservations may lead to problems, as demonstrated in the case of ϵ-overallocation (see, e.g., [134]).

A fluid flow simulation experiment is presented in [79], in which estimates of each connection's future load are used for online performance control of priority system. These estimates are fed to a rate allocation algorithm that decides on how the available rate will be shared among low-priority queues. The objective of this algorithm is to minimize packet loss by keeping all low-priority queues at the same size. The estimates are delivered by an adaptive linear filter, acting on time slots of 100 ms. Also, the influence of the number of coefficients of the filter, i.e., window size or time horizon, on the obtained predictions is discussed.

3.5.5 Traffic Control

Examples on how measurement results are used for the on-line optimization of traffic control parameters are, e.g., found in [18] and [44]. As each domain in general alters the properties of a data stream passing through it, cf.also [183], conforming packets might turn into non-conforming ones. Such packets might be policed in some way, e.g., buffered or thrown out, even though the original traffic profile was in line with the negotiated one. On this background, [44] addresses the recalculation of traffic descriptors, especially the token bucket size, based on measurements of packet-delay variation within each domain.

Chapter 4
Wireless Networks

Wojciech Burakowski
Warsaw University of Technology, Poland

Andrzej Beben
Warsaw University of Technology, Poland

Contributors:

Samuli Aalto (Helsinki University of Technology, Finland), Hans van den Berg (TNO Telecom, Netherlands), Richard Boucherie (University of Twente, Netherlands), Llorenç Cerdà (Polytechnic University of Catalonia, Spain), Markus Fiedler (Blekinge Institute of Technology, Sweden), Roland de Haan (Telecommunications Institute, Portugal), Tobias Hoßfeld (University of Würzburg, Germany), Esa Hyytiä (Helsinki University of Technology, Finland), Sándor Imre (Budapest University of Technology and Economics, Hungary), Lennart Isaksson (Blekinge Institute of Technology, Sweden), Jorma Kilpi (VTT Information Technology, Finland), Remco Litjens (TNO Telecom, Netherlands), Michela Meo (Politecnico di Torino, Italy), Mihael Mohorčič (Josef Stefan Institute, Slovenia), Rastin Pries (University of Würzburg, Germany), Frank Roijers (TNO Telecom, Netherlands), Dirk Staehle (University of Würzburg, Germany), Aleš Švigelj (Josef Stefan Institute, Slovenia), Máté Szalay (Budapest University of Technology and Economics, Hungary), Andrea de Vendictis (University of Roma "La Sapienza"), Branka Zovko-Cihlar (University of Zagreb, Croatia)

4.1 Introduction

The Public Switched Telephone Network (PSTN) and the Internet are indispensable elements of today's life, and wireless mobile access is fast becoming the preferred way for people to connect to these networks. The tendency for using mobile rather than fixed phones is revealed by statistics showing that the number of mobile users is growing each year and will, in the near future, overrun the number of fixed users. A similar trend is observed in the access to the Internet, for which users prefer light mobile terminals, like laptops and PDAs, to fixed terminals.

J. Brazio et al. (eds.), Analysis and Design of Advanced Multiservice Networks Supporting Mobility, Multimedia, and Internetworking, 115–148.

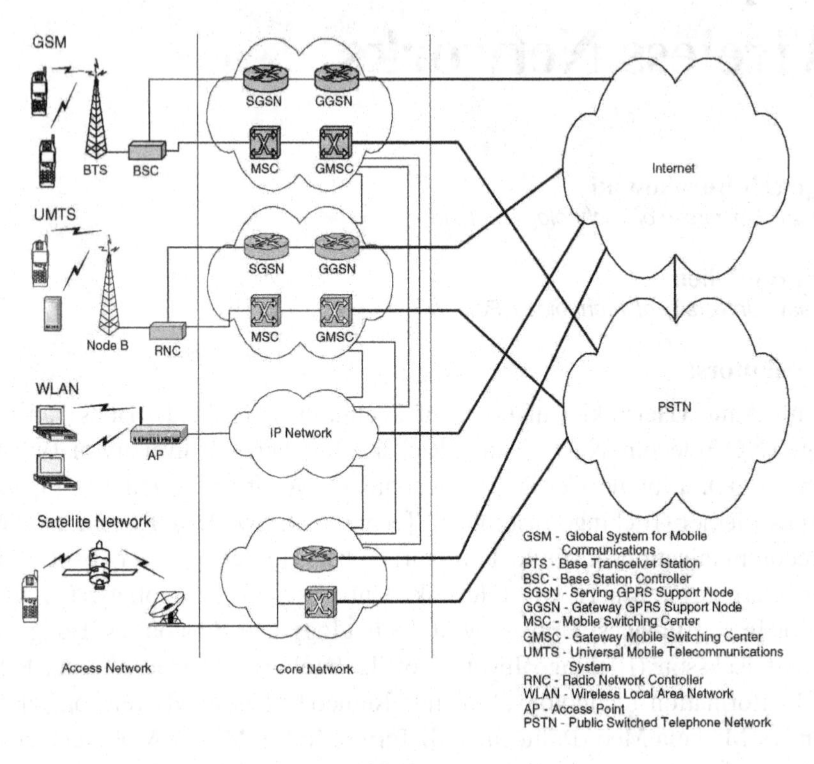

Figure 4.1: Wireless access networks

Wireless access can be provided by a variety of network technologies. Figure 4.1 shows the main technologies in use today: Global System for Mobile (GSM), Universal Mobile Telecommunications System (UMTS), Wireless Local Area Network (WLAN), and Satellite. The increased convenience brought by wireless access should however not mean a decrease in the Quality of Service (QoS) seen by the users as compared to fixed access. For example, the Internet users expect from the service providers new services with interesting content, like news, local info, positioning info, and TV movies. To meet these expectations, new capabilities are required from the network beyond the basic wireless access functionality, in particular, higher capacity of access links and the provision of end-to-end QoS guarantees.

The above expectations have stimulated intensive research and development efforts in the area of wireless and mobile networks. The assumed direction of the evolution scenario is to substitute the existing Second Generation (2G) GSM systems by Third Generation (3G) systems, such as the UMTS and Code Division Multiple Access (CDMA). The 3G systems provide data rates of at least 144 kbit/s for outdoor (mobile) links and 2 Mbit/s for indoor links, considerable higher than those offered by 2G systems, for instance only 9.6 kbit/s in GSM. On the other hand, the WLAN technology, at the beginning treated as an extension of LANs, has recently experienced a phenomenal growth. Currently, the most popular 802.11b WLAN standard offers wireless access link rates of up to 54 Mbit/s, with usage ranging from the office environment to public hot spots. Roughly comparing WLAN and 3G systems, the WLANs have limited coverage but much higher link bit rates, while 3G systems have wider range and provide better mobility support. The wish to have the best of both worlds brings market pressure for the development of solutions allowing users to make seamless handovers between those two types of networks. In this context, and in situations where an appropriately equipped terminal has simultaneous access to more than one wireless network, e.g., WLAN and UMTS, the issue arises of how to set up the connection most suitable to each specific application demand. For solving this problem, the concept of Always Best Connected (ABC) has been developed and is under investigation.

It should be noticed that the wireless links used for access are still of relatively lower bit rates in comparison to the wired links that are typical for the core. Therefore, offering new services requiring end-to-end QoS guarantees necessitates the development of a new network architecture that covers all network technologies and supports both QoS in the core and the access networks, as well as user mobility. Furthermore, while we can do overprovisioning of the core network in a relatively easy way, we have limited resources for the wireless access networks, and thus for achieving QoS we need to implement specific traffic control mechanisms.

The research activities inside COST Action 279 on wireless and mobile networks correspond mainly to issues such as QoS traffic control, traffic measurement and modeling, performance evaluation, and resource management for GSM, UMTS, WLAN, and satellite technologies. The Bluetooth technology is also investigated. These technologies present essential differences from each other, and as a consequence many proposed solutions are technology-specific. Therefore, the present chapter is mostly structured according to the types of wireless networks. Section 4.2 deals with GSM Networks, and Section 4.3 with UMTS Networks. Section 4.4 covers WLANs. The subjects of Sections 4.5 and 4.6 are Satellite Networks and Bluetooth Technology, respectively. The last two sections deal with problems that are not technology specific, and are dedicated to mobility issues and to data transmission over wireless links, respectively.

4.2 GSM Networks

GSM is a circuit-switched technology originally designed for handling voice calls. The General Packet Radio Service (GPRS) is an extension of GSM offering packet switching capabilities over the GSM channels. The technical specifications of GPRS were made in the 3rd Generation Partnership Project (3GPP) [224], and GPRS is currently offered by many European GSM operators. GPRS uses the spare capacity left out by GSM voice or data calls. By default, GSM calls have priority in the use of radio channels. If available, the channels are reserved only temporarily when the terminal receives or transmits data. By using several time slots from the GSM frame, GPRS allows higher data rates as compared to GSM data calls. These data rates and a quite short connection set-up time make the use of Wireless Application Protocol (WAP) services and Multimedia Messaging Service (MMS) cheaper and faster than by using GSM data calls. Moreover, GPRS supports connections to the Internet.

When considering the performance issues of GPRS, the GSM radio channel is one of the main limiting factors, possessing an intrinsic structural delay and bandwidth variations and packet losses that are hard to deal with. The technical specifications give some indications for QoS, but these features are typically not taken in use.

4.2.1 Traffic Measurements and Modeling

In [188] and [202] are presented results obtained from `tcpdump` traces recorded between a GPRS backbone network and the Internet, but before a Network Address Translation (NAT) is made. The reported measurements are from about half a year after the operator had launched its commercial GPRS service. A GPRS session is defined as all packets, and their time stamps, with the same temporary Internet Protocol (IP) address that the user is given at the time of attachment to the GPRS network.

The main observations are that, during the measurement time, most GPRS sessions started during the working days and in working hours were in general roughly similar to low access speed dial-up sessions, and that GPRS session durations and session volumes seem to have heavy-tailed distributions. Also, the majority of the data transfer occurs typically at the beginning of the session, and the GPRS users typically detach from the GPRS network when they have finished active usage.

In [189] are studied distributional properties of GPRS/GSM session volumes and durations, by the application of a known statistical method of ascertaining about the underlying distribution to the data-analysis of GPRS data introduced in [188] and [202]. The statistical method is based on the concept of maximum correlation, and is used because there are no *a priori* reasons to expect or prefer any particular distribution. The main result is that these data sets exhibit a heavy-tailed nature. The Weibull distribution with the shape parameter between $0 < \alpha < 2$ is one of the possible models.

The performance of the Transmission Control Protocol (TCP) over GPRS service was studied in [225]. The authors propose an algorithm to identify TCP spurious retransmission timeouts by post processing of packet traces monitored in an operational GPRS network. The operational principles of the algorithm and the assumptions behind its design are explained in detail, as well as the situations in which the algorithm is prone to inaccuracies. By extensive measurements in a lab testbed using realistic Round-Trip Time (RTT) characteristics and File Transfer Protocol (FTP)-like as well as Web-like traffic generators, it is shown that the algorithm is sufficiently accurate in the detection of spurious retransmission timeouts. Subsequently, the algorithm is applied to real traffic traces captured at the Gi interfaces of an operational GPRS network, to analyse the frequency of spurious retransmission timeouts and the spurious timeout probability as function of the load situation in the network and the flow size. The main finding of this investigation is that spurious timeouts are infrequent events in the considered GPRS network. Finally, by testbed experiments, the effect of RTT variations as seen in the GPRS network is investigated on several flavors of TCP, including New Reno, Selective Acknowledgement (SACK), TCP with Time Stamp enabled, and Forward-Retransmission Timeout Recovery (F-RTO) TCP. The latter protocol is designed to avoid performance degradation due to spurious timeouts.

4.2.2 Resource Management

Resource management policies in GSM/GPRS cellular networks are studied in [226]. Let us recall that these systems basically offer two services: mobile telephony and wireless access to the Internet. Therefore, the resource allocation policy should be carefully chosen so that telephony, from which operators are still getting most of their revenues, is not penalized and, at the same time, data service is provided with a good quality.

Three different resource allocation policies are investigated. According to the *voice priority* channel allocation strategy, priority is given to voice in the access to radio channels. The second channel allocation policy is called *R-reservation*: it statically reserves a fixed number of channels to data services. Finally, the *dynamic reservation* strategy allocates channels to data whenever necessary, using the information about the queue length of GPRS data units within the base station. A threshold on the queue length is used in order to decide when channels must be allocated to data.

The schemes are analyzed by means of Markov chain models. From the results it is shown that voice priority cannot provide acceptable performance to data service, since whenever all the available channels are busy with voice connections, the data service undergoes service interruption. The R-reservation channel allocation policy overcomes this problem and drastically improves the performance of data service. The drawback of this scheme is that it subtracts resources from voice users, even when these are not needed for data, thus inducing an unnecessary performance degradation for voice services. The dynamic reservation scheme provides effective performance tradeoffs for data and voice services, with the additional advantage of being easily managed through the setting of the threshold value.

In [227] are considered adaptive dynamic channel borrowing strategies for wireless networks covering a road. In a Fixed Time-Division Multiple Access (FTDMA)-based network model, road traffic prediction models are used to characterize the movement of hot spots, such as traffic jams, and subsequently to predict the traffic load offered to the network. A dynamic upper bound on the capacity required to achieve a specified QoS level in the cells is computed. Restricting borrowing to neighboring cells, to avoid excessive re-allocation of capacity, optimal channel borrowing strategies based on traffic movement and traffic density are given. These strategies can be characterized by a straightforward rule of thumb, of easy implementation: borrow capacity from the cell on the steeper side of the traffic peak. The simulation results under realistic traffic load conditions indicate a significant reduction of call blocking probabilities assuming the proposed optimal channel borrowing strategy.

The problem of load balancing in cellular networks is discussed in [228]. The performance of both static and dynamic call routing policies for a simple model of two base stations with overlapping cells is compared. The new call originating from the overlapping area can be routed to either one of the two base stations. In a static policy the routing decisions are based only on the system parameters, i.e., arrival rates and mean holding times, assumed to be known, whereas a dynamic policy uses additional information about the system state, defined by the number of running calls in each station. Since the capacity of the base stations is finite, it may happen that a new call will be blocked. A method for constructing a reasonable routing policy, close to the optimal, is proposed.

The approach is based on the theory of Markov Decision Processes (MDP). The objective is to minimize the call blocking probability. In particular, the blocking performance of the policy obtained by the First Step of the Policy Iteration (FPI) algorithm is investigated. Under a randomized policy the two base stations can be modeled independently as Erlang loss systems, for which it is easy to determine the so called Howard's relative costs in each state. This makes the FPI of low complexity, and therefore applicable for large instances. In addition, it is easy to determine the optimal policy among the randomized policies, which is a subset of the set of static policies.

The idea is to approximate the optimal dynamic policy with the FPI policy. It is shown that any FPI policy fulfils two reasonable requirements: it is greedy, and of the threshold type. In addition, numerical experiments for small instances provide a deeper understanding of the FPI policy. Interestingly, it turns out that starting with the optimal randomized policy as the basic policy does not necessarily lead to the best performance of the FPI policy. Therefore, a heuristic rule is suggested for the basic policy. With such a choice, the FPI policy seems to be close to the optimal dynamic policy, and performs better than the other considered policies.

4.3 UMTS Networks

UMTS is the 3G wireless network standard for providing different QoS services and operating with bit rates up to 2 Mbit/s. This is achieved by operation with Wideband Code Division Multiple Access (WCDMA) over the air interface.

Given the fundamental differences from GSM, the introduction of UMTS requires new paradigms in wireless network design. Recall that in GSM the capacity of a base station is determined only by the number of available frequencies, and hence is independent of current network load. Given the number of frequencies available in a cell, the allowed network load follows directly from the Erlang-B formula, since the GSM network provides mainly voice calls. In contrast, the capacity of a Base Station (BS) in a WCDMA network is interference limited. On the uplink direction, the Multiple Access Interference (MAI) at a BS is caused by all the Mobile Stations (MS) both from the given BS and from neighboring BSs. On the downlink direction, the capacity is limited by the transmit power of the BS or by the interference level. The power control mechanisms in both link directions control the transmitted powers in such a way that, for each service, signals are received with nearly equal strength. A detailed examination of the interference on the uplink of a given cell is no straightforward task. Due to the universal frequency reuse in UMTS, all users, both in the considered cell and in the neighboring cells, contribute to the total interference, thus influencing the link quality in terms of received bit-energy-to-noise ratio (E_b/N_0).

The planning of WCDMA networks consists of two aspects: the coverage planning and the capacity planning. In contrast to GSM, the coverage and the capacity cannot be considered as independent terms. In WCDMA, a trade-off between the coverage area and the capacity of a BS exists. The more users are active at a BS, the larger is the MAI at the BS, and the higher are the transmit powers required by the MSs to fulfil their E_b/N_0 requirements. Additionally, due to the restriction of the MSs transmit power, the coverage area shrinks with an increasing number of users. Attaining a certain coverage area for a BS demands a limitation of the MAI, which can be done by admission control. The MAI level used as threshold for the acceptance of new calls determines not only the coverage area, but also the capacity of the BS.

Another difference between GSM and WCDMA is the handover procedure. While GSM supports only hard handovers, where the connection to the new BS is established after terminating the one to the old BS ("break before make"), *soft handover* is performed in WCDMA. Here, the mobile assists in the handover process by measuring the pilot signals from the neighboring BSs and storing those BSs with the strongest received signals in the Active Set (AS). The mobile then communicates with all BSs in the AS simultaneously ("make before break"). As a consequence, the MS receives multiple power control commands and adapts its transmission power on the uplink to the BS with the least requirement.

The introduction of 3G mobile communication systems also allows the service providers to offer a large variety of services, categorized in UMTS under the classes conversational, streaming, interactive, and background. While the conversational and streaming classes have a guaranteed bandwidth and delay, the interactive and background (best effort) classes consume the remaining system capacity. On the downlink this system capacity is limited by the BS transmit power, and on the uplink by the interference. As traffic in UMTS networks is expected to be asymmetric, with the bulk of it towards the MS, the downlink becomes the limiting link for best-effort traffic. In UMTS, best-effort traffic may be carried either on dedicated channels that are subject to rate control or on a shared channel. UMTS release 5 further standardizes the High-Speed Downlink Shared Channel (HS-DSCH), which achieves data rates up to 10 Mbit/s.

4.3.1 Admission Control for WCDMA Systems

The key feature of WCDMA systems is that all users transmit in the same frequency band, their signals being separated by the use of orthogonal or pseudo-orthogonal codes. Except for the ideal case, when real orthogonal codes are used and no multi-path propagation occurs, a user sees the other users' signal as interference. The total interference comprises the own-cell interference $\hat{I}_{own}$, the other-cell interference $\hat{I}_{other}$, and also the thermal noise $\hat{N}_0$. The interference grows with the number of calls in progress and limits the capacity.

WCDMA Admission Control (AC) is performed on the basis of the measured *noise rise*, defined as the ratio of the total interference $\hat{I}_0$ to the interference $\hat{N}_0$ of an unloaded (empty) system. The AC estimates the increase of the noise rise that would be caused by accepting a new connection, and blocks it if the result exceeds a predefined threshold. While the noise rise is a value that is measured by a BS, it is not well suited for understanding the actual system load. A transformation of the noise rise yields the definition of the cell load η:

$$\text{Noise rise} = \frac{\hat{I}_0}{\hat{N}_0} = \frac{1}{\frac{\hat{N}_0}{\hat{I}_{own}+\hat{I}_{other}+\hat{N}_0}} = \frac{1}{1 - \frac{\hat{I}_{own}+\hat{I}_{other}}{\hat{I}_{own}+\hat{I}_{other}+\hat{N}_0}} = \frac{1}{1-\eta} \quad (4.1)$$

A cell load equal to 1 defines the pole capacity of a WCDMA cell. On the arrival of new call submitted to service t, the AC algorithm estimates the additional load α_t brought in by the call. This load is based on the negotiated traffic contract parameters, i.e., bit rate and maximum error rates. WCDMA AC consequently accepts an incoming connection if the estimated cell load η_{est} stays below the predefined threshold value η_{max}. The acceptance of a new call depends on both the own-cell interference and the other-cell interference. This feature explains why we speak of soft-blocking in WCDMA networks.

A time-efficient algorithm to compute blocking probabilities in a WCDMA network operating with several services is proposed in [229] under the assumption that users submit calls to each service class according to a Poisson process with exponentially distributed holding times. The load produced by a user is declared by the submitted QoS requirements, i.e., bit rate and target E_b/N_0, and is called load per service. The user activity at an arrival instant is modeled by a Bernoulli random variable. The AC condition which originally relates to own- and other-cell interferences is transformed into own-cell load η_{own} and other-cell load η_{other}. The other-cell load is modeled as a lognormal random variable that is assumed to be independent of the own-cell load. This allows the derivation of the probability $\beta_t(\eta_{own})$ that a call of service t is blocked in a system state with own-cell load η_{own}. This probability is called the soft blocking probability and can occur in virtually every state, depending on the other-cell load.

Figure 4.2 shows an example of the blocking probabilities for a system operating with three services. The dotted lines show the approximated blocking probabilities and the solid lines correspond to simulation. One can see that the approximation yields accurate results in the range of scenarios presented, possessing strongly varying other-cell interference levels and different

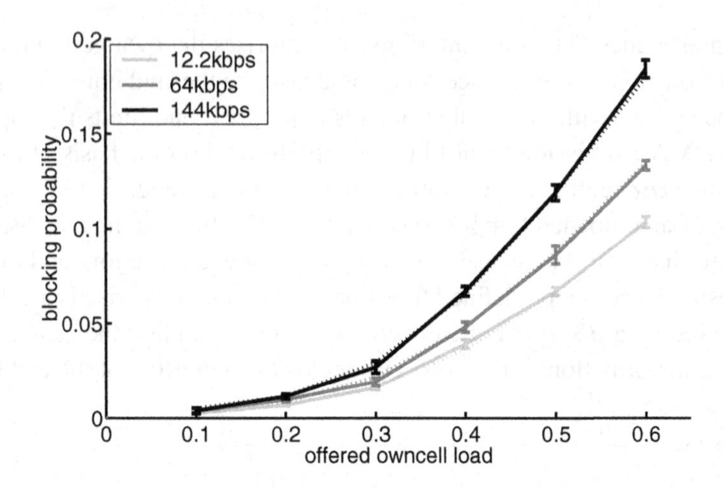

Figure 4.2: Uplink blocking probabilities for three services with 12.2 kbps, 64 kbps, and 144 kbps

user activities of 0.45, 0.3, and 0.8 for the three services. These are the especially interesting cases, since with deterministic other-cell interference and Always-ON users the proposed analysis yields exact results.

The main intention of [230] is to show that we can apply well-studied methods for analyzing the UMTS capacity to web traffic. The focus is on the QoS web service, in particular, on the number of QoS web users that a NodeB is able to handle in parallel. The capacity analysis of the uplink and the downlink of CDMA and WCMDA systems mainly assume a Poisson distributed number of users per cell and service. A service is defined by its bit rate, target E_b/N_0, and activity factor or mean activity during its sojourn time in the system. The paper first shows by means of a detailed simulation that it is possible to describe a QoS web service by these parameters. This detailed simulation includes a sophisticated web traffic model, the implementation of TCP according to 4.4BSD-Lite, and the power control according to the 3GPP standard. As a result of this simulation is obtained the distribution of the number of active web users, where "active" means actively transmitting on the UMTS downlink, and the distribution of the NodeB transmit power. A second simulation that does not consider power control is used to obtain the probability distribution of web session durations and web session activities. In the first instance, this distribution is used in an activity simulation to describe the web sessions and show that the results match with the detailed simulation. It is thus shown that the Poisson assumption holds for the number of web users that simultaneously transmit over the air interface, and that the resulting NodeB transmit power distribution is valid. In a further step, an offered load from the web session duration and web activity distribution are calculated, and are used in the Monte Carlo simulation to determine the NodeB transmit power distribution according to a certain offered web traffic load. In this way, the web capacity of an example UMTS network can be evaluated.

An alternative way to perform AC in WCDMA networks is discussed in [231]. The AC algorithm combines the dynamic optimization of the Chernoff bound instead of the well-known static effective bandwidth concept. A very important advantage of the proposed method is its dynamic behavior that, in contrast to the traditional static effective bandwidth methods, allows resilient adaptation to the continuously changing network parameters. The proposed algorithm is able to adapt dynamically to an ever-changing radio environment, and provides a trade-off between decision efficiency and complexity. The proposed AC method is investigated under ON/OFF traffic sources and lognormal fading channels.

The problem of call assignments to base stations on downlink in a heterogeneously loaded, linear UMTS network is analyzed in [232]. The investigated model covers the situation along a highway where, due to traffic jams ("hot spots"), the load of the cells is not distributed evenly along the road. By dividing the area into small segments, the power requirements are characterized via a matrix representation that separates user and system characteristics. A closed-form expression of the Perron-Frobenius (PF) eigenvalue of that matrix is obtained, providing a quick assessment of the feasibility for each distribution of calls over segments (and for each assignment of calls to base stations). In particular, the PF eigenvalue is almost linear in the number of calls per segment, and thus provides a kind of "effective interference characterization of downlink feasibility". The results allow for a fast evaluation of outage and blocking probabilities, and may be used for call acceptance control.

Evaluation of interference aspects of direct sequence spread spectrum mobile communication systems is the subject of [233]. Interference conditions in both the uplink and downlink transmission are discussed. For calculation of interference conditions, a statistical method based on the average emission power of a mobile station in a cell is proposed, and compared with one based on calculation of multiple cell interference reduction factors. The study shows that these two methods produce similar results.

4.3.2 Soft Handover

In CDMA systems, during the soft handover process MSs are connected, not only to one, but to several BSs. An MS moving in an area with several BSs has

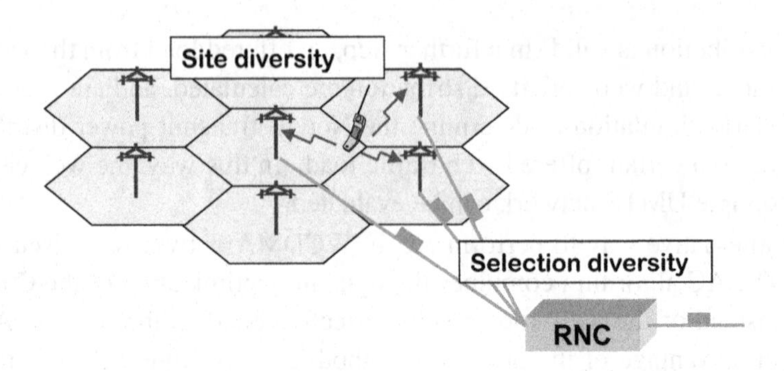

Figure 4.3: Diversity effects due to soft handover in WCDMA networks

an AS that changes dynamically and is determined by the pilot signal transmitted by every BS. An MS detects the BS with the strongest received pilot signal and also those BSs with a signal strength not more than the *reporting range* below the strongest signal. All these BSs form the Active Set of an MS.

On the uplink, all BSs in the Active Set receive the frames transmitted by the MS (*site diversity*) and transfer them to the Radio Network Controller (RNC). There, all frames are checked for errors, and only if all of them are erroneous does a frame error occur (*selection diversity*), see Figure 4.3. The RNC evaluates the resulting frame error rate and adapts the target E_b/N_0 in the outer loop power control. This target E_b/N_0 is signaled to all BSs in the Active Set, which try to adjust the transmission power of the MS to this value according to the inner loop power control. The MS receives power control signals from all BSs in the Active Set and increases its power only if all BSs signal *power up*. Otherwise, if one or more BSs signal *power down*, the MS obeys the latter command. On the uplink, soft handover leads to a reduction of the required transmission powers, and consequently to less interference in the system, from which an increased system capacity results. The benefits of soft handover on the uplink include automatic load balancing, a target E_b/N_0 reduction, and increased robustness against fading. The automatic load balancing and the robustness against fading result from the inner loop power control, which "selects" the best BS on slot level, i.e., 15 times per 10ms interval. The best BS is not solely determined by the propagation loss, but also depends on the interference levels at the BSs in the Active Set. MSs at the border of a highly loaded BS may be power-controlled by a nearby BS with less load, even though the propagation loss to this BS is higher. Thus, the load is shifted from BSs with high load to BSs with lower load. This effect is investigated in [234] using Monte Carlo simulation techniques. A series of snapshots of a UMTS network with 39 BSs in a hexagonal grid is generated for homogeneous and non-homogeneous spatial traffic distributions. The soft handover gain, defined as the reduction of the mean interference due to soft handover, grows in the homogeneous scenario with the offered load, and reaches a maximum of about 3dB. In non-homogeneous scenarios, even higher soft handover gains are obtained, since there is more potential for load balancing. The outer loop power control further increases the soft handover gain by decreasing the target E_b/N_0 value, such that lower received powers are required at the BSs. This gain is additional to the benefit obtained by selecting the best BS, as still all BSs in the Active Set try to adjust the MSs transmit powers to the same target value. The effect of the outer loop power control is investigated in [235]. In this paper, soft handover gains up to 8dB are reported in homogeneous networks with high traffic density.

4.3.3 Handling of Packet Data Traffic in UMTS Networks

Wireless data transfer is indisputably a major driver for the deployment and anticipated success of 3G mobile networks. As the foreseen (data) services greatly differ in their traffic characteristics and QoS requirements, a number of distinct transport channels have been specified to accommodate these services efficiently. In the downlink, commonly the focus of the attention in light of the generally expected strong up/downlink data traffic asymmetry, the Dedicated Channel (DCH), the Forward Access Channel (FACH), and the Downlink Shared Channel (DSCH) are standardized. Specifically designed for delay-sensitive services or services with stringent throughput requirements, the DCH is a bit pipe assigned exclusively to a single mobile station, and has the advantage of fast closed-loop power control and macro-diversity. On the other hand, the FACH and DSCH may be shared by multiple mobiles. The FACH is typically used for the transfer of relatively small data chunks, without the advantages of closed-loop power control and macro-diversity. Medium to large data transfers, particularly of bursty character, e.g. TCP/IP flows, are most efficiently conveyed on the DSCH, as it enjoys the advantages of closed-loop power control by maintaining a low bit rate Associated DCH (A-DCH) for signaling purposes. A principal advantage is the enhanced efficiency of channelization code usage. Since data is multiplexed on the DSCHs, the use of soft handover, or macro-diversity, is rather complicated from an implementation viewpoint and therefore not standardized. UMTS release 5 further standardizes the High-Speed Downlink Shared Channel (HS-DSCH), which achieves data rates up to 10Mbit/s by a variety of enhanced technologies including (i) higher order modulation together with fast link adaptation, i.e., adaptive coding and modulation, optimized for channel conditions, (ii) fast scheduling centered at the Node-B rather than at the RNC, with a proposed smaller transmission time interval, in order to reduce delays and facilitate better tracking of the channel variations, e.g. equal to the 0.67ms slot duration, and (iii) fast cell selection, in order for an MS to continuously select the serving cell with the best radio conditions, important in light of the unavailability of macro-diversity on the HS-DSCH.

In [143] is presented a semi-analytical performance evaluation of the DSCH, investigating the influence of A-DCHs and interference from neighboring cells on the downlink transfer rates. The analysis of a UMTS network with N cells consists of two stages. In the first stage, the cell-wise outage probabilities for a given number of calls per cell are obtained using Monte Carlo simulation. The second stage captures the traffic dynamics of call arrivals and terminations in an N-dimensional irreducible continuous-time Markov chain. A system state is specified by the number of calls per cell, and the outage probabilities obtained in the first stage are integrated by reduced arrival rates. Furthermore, for each state and cell an effective throughput per call is determined as function of the frame error rate.

The influence of A-DCHs and other-cell interference is demonstrated by simple experiments. The data transfer rates of networks with one, two, and three Node-Bs are evaluated with and without A-DCHs. The experiments show that a heavier data traffic load implies both a greater competition for DSCH resources, and thus longer transfer delays, and a higher interference level, due to the greater number of A-DCHs that must be maintained for signalling purposes. This latter effect causes a higher frame error rate, and thus a lower effective aggregate DSCH throughput. Hence, the greater the demand for service, the smaller the aggregate service capacity.

In [236] is investigated the performance of an integrated services UMTS network with speech calls on DCHs and data calls on an HS-DSCH. Different scheduling schemes for micro- and macro-scheduling are compared. *Macro-scheduling* adjusts the power of the HS-DSCH to the varying traffic conditions at a time scale of seconds and has inter-cellular scope. *Micro-scheduling* time-multiplexes the data flows within each cell in order to optimize resource efficiency according to varying channel conditions, while satisfying the call's QoS requirements and providing some sort of fairness. Two different scheduling disciplines are considered for micro-scheduling: *power-fairness* and *rate-fairness*. Power-fairness means that within one cell the same power is spent for all users, and thus users with worse channel conditions experience a lower rate. Rate-fairness means that all users within one cell obtain the same rate, but more power is spent for users with worse channel conditions. Macro-Scheduling can be either *adaptive* or *fixed*. Adaptive means that the power allocated to the HS-DSCHs of the different cells optimizes the data throughput while maintaining the C/I requirements of the speech users.

The different disciplines for micro- and macro-scheduling are compared using Monte Carlo simulation techniques. For a series of snapshots, the outage probability for speech calls and the achieved rate for data users are evaluated as performance criteria. The principal objective of the experiments is to demonstrate the potential performance enhancements of adaptive macro-scheduling. The results show that indeed adaptive scheduling leads to an improved QoS for both speech and data services. Speech calls enjoy the most significant performance gain due to the strict priorization strategy. For the fixed macro-scheduling scheme, there is a trade-off between the QoS for speech and data users. More power for the HS-DSCH leads to a higher transfer rate for data users, but also to a higher outage probability for speech users. Power-fair micro-scheduling leads to a higher expected throughput per data call than rate-fair micro-scheduling. However, the expected throughput for users at the cell border is significantly smaller than for users near the center of a cell.

Performance of High-Speed Downlink Packet Access

In [145] is presented a performance evaluation for a UMTS High-Speed Downlink Packet Access (HSDPA) network, considering the key HSDPA mechanisms of Adaptive Modulation Coding (AMC), channel-aware scheduling, and hybrid Automatic Repeat Request (ARQ) in a more accurate sense compared to [236]. In contrast to the assumption of (semi-)persistent data flows commonly found in related literature, the investigation concentrates on incorporating the flow level dynamics in the evaluation models, which refers to the initiation and completion of finite flows at various locations, leading to a varying number of concurrent flows competing for shared resources. New qualitative and quantitative insights into the HSDPA flow level performance are provided by decomposing the performance with respect to the relative impact of some key system, environment and traffic-related aspects. Particular focus is placed on the impact of terminal location, the presence of multipath fading and intercellular interference, the inherent feedback delay in the channel quality reports, the correctional capabilities of the hybrid ARQ scheme (soft combining), the flow level traffic dynamics, the flow size variability, and the applied packet scheduling scheme. The contribution of these key aspects in the experienced service quality and spatial performance fairness is assessed by evaluating gradually more 'complete' scenarios. The performance evaluations are conducted using dynamic simulations and stochastic analyses when feasible, e.g. multi-class Processor Sharing (PS) and priority queueing models. Among the insights gained, the results indicate that the presence of multipath fading has a positive effect on the flow level performance, in particular under channel-aware schedulers. The channel quality feedback delay causes a severe performance degradation, even for low fading velocities. The increased number of block errors due to the feedback delay can be partially coped with by the hybrid ARQ (soft combining). Overall, for the considered settings, the pure Signal-to-Noise (SNR)-based scheduler outperforms the other considered schedulers, including the well known proportional fair scheduler, with respect to the transfer time performance, even for data flows near the cell boundaries, which is in contrast to the claims reported in other studies that disregard the significant impact of the flow level dynamics.

In [146] the focus is on the spatial throughput unfairness in UMTS/HSDPA networks, that is inherently induced by HSDPA's AMC and channel-aware scheduling mechanisms. The former effect is that remote data flows generally experience worse radio link qualities, are therefore served with lower data transfer rates, thus experiencing longer holding times, and hence are found in greater traffic relative densities when observing the system at a random time instant. The second effect is that under channel-aware scheduling, remote data flows tend to be scheduled for transfer less frequently, with a similar effect on the flow holding time and the spatial traffic distribution. The above-mentioned impact of the AMC and scheduling schemes on the spatial traffic distribution is assessed by means of both dynamic simulations and stochastic analyses, for a range of data traffic loads, indicating that the 'AMC effect' is larger than the 'scheduling effect'. UMTS network planning tools generally rely on snapshot-based performance evaluations, which are inherently characterized by a number of drawbacks, e.g. the lack of consideration of the endogenously affected spatial data traffic distribution and the limitation to time- rather than to the more relevant flow-centric, performance measures. In this light, the paper further investigates both drawbacks. In particular, the realistic flow-average throughput values obtained via dynamic simulations are compared with the time-average throughput values resulting from snapshot-based simulations. Among other results, the comparison reveals that the errors of considering time- rather than flow-average throughput performance, and an erroneous yet fairly natural default assumption of spatial traffic homogeneity, appear to be in opposite directions and for the considered parameter settings even roughly cancel out. Although this observation seems fortunate, the significance of each individual discrepancy is sufficiently large to motivate a more extensive numerical assessment.

4.3.4 UMTS Network Planning

The mobility of terminals in a UMTS network strongly influences network capacity. On the one hand, this impact stems from the more severe Signal-to-Interference-Ratio (SIR) requirements that apply in case of higher velocities due to the combined effects of multipath propagation, Doppler shifts, and power control imperfections. On the other hand, an increased mobility requires a higher level of radio resource reservation regarding handovers, in order to keep the call dropping probability below a prespecified target value. As a consequence, fresh call blocking increases, inducing a need for denser site planning. In [237], an analytical approach is presented to evaluate the impact of these two mobility-related aspects on network planning and performance, and on investment costs. The principal strength of the approach lies in the model being simple enough to allow a computationally relatively inexpensive performance evaluation and optimization, yet being sufficiently realistic to provide valuable qualitative insight for network planning purposes.

The fresh call blocking probabilities and call dropping probabilities are evaluated using a two-dimensional time-continuous Markov chain with a state described by the number of users in the considered cell and in the surrounding cells. The interference of a user in a surrounding cell is modeled by a Gaussian random variable including terminal location and call activity. By demanding a given outage probability, the soft capacity of a WCDMA cell is transformed into a fixed set of accessible states. The velocity of the users is incorporated into the different SIR requirements and cell residence times. Furthermore, the call admission control for fresh calls reserves a certain radio resource for handover calls. The amount of reserved radio resource and the cell radius are optimized in order to minimize network investment costs in terms of BSs per covered area, while meeting the target call blocking and dropping probabilities. The primary conclusions from the numerical examples are that (i) the impact of terminal velocity on the optimal cell radius, and therefore on the investment costs, can be quite significant, potentially up to a factor of 2, (ii) the deployment of a radio resource reservation scheme can indeed be effectively utilized to reduce call dropping and investments costs, and (iii) planning a UMTS network using inaccurate estimates of terminal velocity can lead to unacceptable blocking and, in particular, dropping probabilities.

In [237], the influence of mobility on the radio resource reservation for handover calls is demonstrated using a simple mobility model, with cell residence times with exponential distribution, and expectation dependent on the mobile's velocity.

In [238], a new mobility model is introduced in order to predict the number of terminals in each cell and to improve the effectiveness of Connection

Admission Control (CAC) algorithms. This model is studied in the context of other mobility models reported in the literature, namely random walks and Markov models.

Two-dimensional Markov models are very complex, because of the need to keep state for the six adjacent cell directions. The goal to define a simple yet appropriate model is achieved by extending the one-dimensional model to a two-dimensional model that retains its simplicity by limiting the possible states of the mobile user. The main idea of the model is to separate the neighboring cells into two groups according to the typical user movement direction. The number of mobile terminals for future time periods is calculated relying on the random walk model and a modified Markov mobility model. The calculation of the future number of mobile terminals in cell *k* uses the concept of a ring, defined as the cells surrounding cell *k*. This concept simplifies the calculations, because the interest rests only on the number of users arriving to or leaving a given ring during a time period, whereas internal movements (inside the rings) remain unconcerned. The goal is to predict the number of users in cell *k* in time slot t_{i+1} and t_{i+2}, based on the number of users in cell *k*, the first ring, and the second ring at time slot t_i. The accuracy of different mobility models is compared by simulation. The results and equations obtained can be utilized in resource-reservation based AC algorithms, paging algorithms, etc. The parameter of the random walk model is easily determined from measurements of the actual number of handovers in the network. The parameters of the modified Markov model can also be calculated from handover measurements. In addition, the directions of the handover events and the time interval between the handover events are needed. Predictions based on the extended Markov model prove to be as precise as the handover vector random walk method, but the required calculations and equations are much simpler. The results show that the accuracy of the prediction depends on the range of the forecast. The two-ring concept takes into consideration the users moving at higher speed, therefore provides more accurate information for the AC algorithm than the ring based forecast. It is left open for future work the determination on the optional forecast distance as function of the mobile terminal's speed in the network.

4.4 Wireless LANs

Recently, WLANs have gained a prominent role in the telecommunications environment because of their support of seamless access of portable devices, like laptops and electronic organizers, to the Internet. As the Internet becomes a multi-service network, one can also expect from WLANs the internal capabilities for providing guarantees on some QoS Parameters, such as delays and packet loss. The WLAN standards, like HIPERLAN and IEEE 802.11, mainly concentrate on the physical, data link, and Medium Access Control (MAC) layers. However, for supporting user mobility in a wide area (outdoor), enhancements at the network layer such as mobile IP are required. Physical layer issues need to be taken into account because interference caused by other systems (WLANs, Bluetooth) or by noise strongly influences the available capacity of the system. From a QoS point of view, efficient scalable MAC protocols that can handle traffic with different QoS requirements are expected. The IEEE 802.11e standard [239] attempts to fulfil these demands, but only with moderate success. In addition, roaming and handover have to be considered, since a significant amount of delay and packet loss can be traced back to these events.

The activities of the COST Action 279 in the area of WLAN networks have mostly focused on analytical models for performance evaluation, reporting measurements results, methods for improving traffic handling, and handover issues.

4.4.1 Analytical Models

An integrated packet/flow level model suitable for analyzing flow throughputs and transfer times in IEEE 802.11 WLANs is proposed in [144]. The packet level model captures the statistical characteristics of the transmission of individual packets at the MAC layer, while the flow level model takes into account the system dynamics due to the initiation and completion of data flow transfers. The model is a PS type of queuing model, reflecting the IEEE 802.11 MAC design principle of distributing the transmission capacity fairly among the active flows. The resulting integrated packet/flow level model is analytically tractable and yields a simple approximation for the throughput and flow transfer time. Extensive simulations show that the approximation is very accurate for a wide range of parameter settings. In addition, the simulation study confirms the attractive property, following from the approximation, that the expected flow transfer delay is insensitive to the flow size distribution, apart from its mean.

In [240] is presented a flow level modeling approach for analysing TCP flow throughputs and transfer times in IEEE 802.11 WLANs. The model captures the behavior of TCP packets sent over the MAC layer and takes into account the system dynamics due to the initiation and completion of data flow transfers. In particular, at the flow level the system is modeled by a PS type of queue, reflecting both the IEEE 802.11 MAC design principle and TCP behavior of sharing the transmission capacity fairly among the active flows, although the TCP behavior is dominant. The modeling results are compared to simulation results and illustrates that the PS model estimates the transfer times well. Only for small files does the PS model underestimate the transfer times, as the TCP slow start behavior is ignored.

In [241] are considered two popular WLAN protocols, Carrier Sense Multiple Access (CSMA) and Ready-to-Send-Clear-to-Send (RTS-CTS), and is studied the extent to which they deviate from an ideal protocol, in the sense of either (i) not allowing transmissions that would be successful, were they allowed, or (ii) allowing transmissions that immediately get destroyed by ongoing transmissions or destroy such transmissions. The paper starts by examining how, in topologies with partial connectivity, the interaction between control packets, data packets, and the protocol rules can give rise to situations where the two types of behavior above can occur, paying special attention to the so-called hidden stations, exposed stations, gagged stations, and masked stations, the latter two types only occurring under RTS-CTS. It then formulates an analytical Markovian model for the determination of link throughput in an environment with an arbitrary pattern of hearing between pairs of stations, having as input variables the rates at which the stations attempt packet transmissions to each of their neighbors. An optimization framework is next developed to allow the determination of the system capacity, which refers to the situation of maximum link throughputs compatible with a given set of pre-specified ratios between pairs of link throughputs. Finally, a configuration of two WLAN cells with inter-cell mobile interference is studied by means of this framework for the purpose of comparison of the system capacity afforded by CSMA, RTS-CTS, and an ideal protocol. The results show that the protocol limitations identified have a significant influence on their capacity, as revealed by a comparison with the ideal protocol. In addition it is seen that, even neglecting the higher overhead of RTS-CTS over CSMA, the former can perform significantly worse than the latter, mostly in situations where CSMA allows successful concurrency of transmissions that are not allowed by the gagged-station problem of RTS-CTS.

4.4.2 Measurements

In [195] are presented the results of experiments carried out in an infra-structured 802.11b WLAN comprising a single Access Point (AP) and a variable number of user terminals. The aim of this work is twofold: on the one hand, well-known "TCP (UDP) over 802.11" analytic models are validated by means of experimental results; on the other hand, the behavior of TCP over 802.11b is analyzed. This study is performed over a highly configurable environment

(Linux operative system, "hostap" driver, etc.), that allows to vary several key system parameters, e.g., Transmission Bit Rate and maximum number of MAC retransmissions. Flow fairness, interaction between WLAN link layer parameters (e.g., ARQ retransmission persistence degree) and transport protocols, and TCP flows traffic characteristics are investigated. As for the main results, the paper

- validates UDP and TCP analytic models by means of accurate experiments; the models seem to capture the behavior of these protocols accurately;
- points out the impact of the "capture effect" on UDP traffic in terms of throughput and fairness: although in some case the "capture effect" improve the cumulative UDP throughput, it strongly reduces the global fairness;
- verifies that TCP preserves fairness even in presence of the "capture effect", in particular when the traffic is in the uplink;
- observes particular TCP traffic patterns involving burstiness, that can also explain the TCP fairness in presence of the "capture effect".

A measurement study of an operative wireless LAN is presented in [177]. The performance of TCP is evaluated in an infrastructured 802.11b LAN with an access point and a number of wireless stations sharing the access link. Besides traditional aggregate performance indices at the link layer, the measured parameters include TCP level variables, such as the estimated segment round-trip time, the maximum achieved TCP congestion window size, and the number of retransmitted segments. ¿From the presented results, a number of observations can be drawn. When most of the traffic flows downlink and the access link is the bottleneck of the TCP connections, some unfairness between connections arises and the performance indices variance is significant. By reducing the maximum congestion window size, such effects can be reduced. In scenarios where most of the traffic flows uplink, the medium access control scheme plays a crucial role and, in this case, the use of RTS/CTS can be beneficial in terms of both total throughput and fairness. Indeed, while in the literature it was previously shown under slightly different scenarios that the RTS/CTS mechanism improves the fairness but reduces the throughput due to the larger overhead, in the considered scenarios, by reducing the TCP segment loss probability, the RTS/CTS mechanism also improves the throughput.

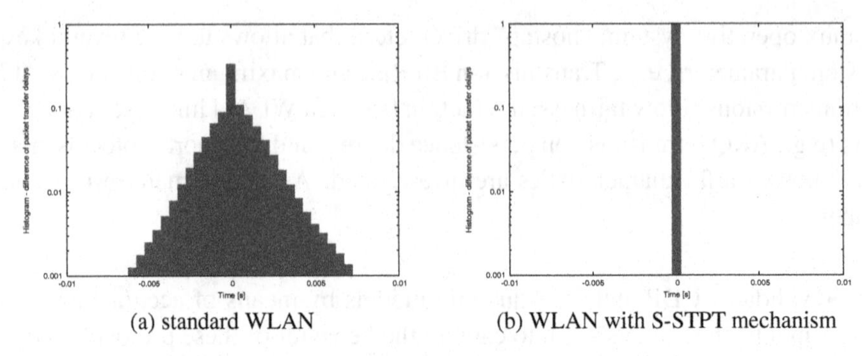

(a) standard WLAN (b) WLAN with S-STPT mechanism

Figure 4.4: Packet transfer delay characteristics

4.4.3 Methods for Improving Traffic Handling

In this section are briefly described approaches proposed for improving the handling of both streaming and elastic traffic in WLANs by adding traffic control mechanisms above the MAC layer. It should be noticed that such direction of research is of special importance since it looks at mechanisms that do not require modification to the MAC layer, and therefore can be relatively easily deployed even in existing systems.

Streaming Traffic

In [242] and [243] is proposed and described a mechanism, implemented on top of the MAC layer and designated Self-Synchronized Time Packet Transmission (S-STPT), for the transmission of Constant Bit Rate (CBR) traffic in WLANs with very low delay variation. The proposed method attempts to limit as far as possible the occurrence of packet collisions, whose associate retransmissions are the main cause of packet delay variation. The main idea of S-STPT is to set the times at which particular terminals send their packets according to the observation of packet conflicts in the wireless channel, delaying the packet transmissions so as to guarantee that no conflicts occur. The effectiveness of the S-STPT mechanism is verified in [242] and [243] for the cases of homogeneous and heterogeneous CBR traffic. In these papers, some implementation issues are also discussed.

Figure 4.4 shows example histograms of the difference between the delays of consecutive packets, i.e., the delay jitter, collected in (a) the standard WLAN 802.11b system, and (b) a WLAN with the S-STPT mechanism. One can observe that with S-STPT the packet delay variation is practically eliminated.

It is worthwhile noticing that, for most of the cases, self-synchronization of the system in achieved after the transmission of only a few packets, as illustrated in Figure 4.5.

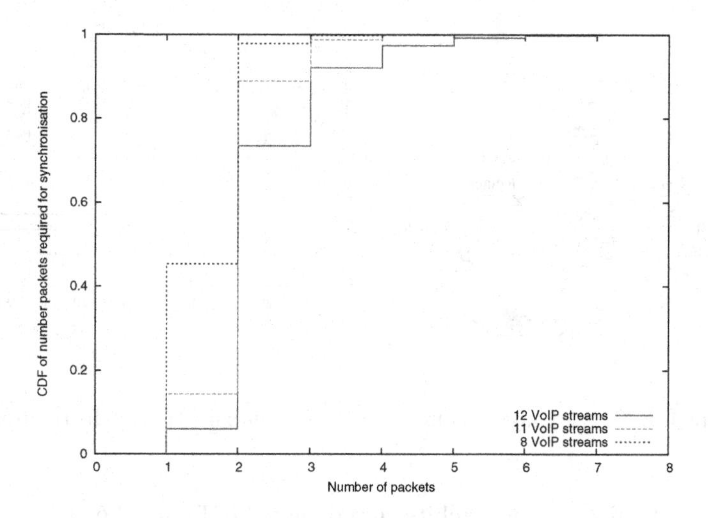

Figure 4.5: Complementary distribution function of time needed for finishing self-synchronization

Elastic Traffic

Despite the increasing popularity of WLANs based on the IEEE 802.11 technology, there is still the need for solutions that guarantee fairness and QoS levels. In [244], a Logic Link Control (LLC)-layer algorithm is proposed that aims at guaranteeing fair access to the medium to every user of a 802.11-based infrastructured WLAN. The proposed scheduling mechanism is based on two features. First, the channel conditions are estimated for each wireless station, so that the stations which undergo bad channel conditions refrain from transmitting until the channel returns to good a condition. Second, by keeping track of the achieved throughput for each connection, longer transmission opportunities can be given to wireless stations that, having experienced channel failures, received lower throughput than their fair share. Compared to the standard solution, the proposed algorithm jointly provides higher network throughput and fair access to the radio resources. Moreover, the proposed scheme can protect short-lived TCP flows that, by being particularly sensitive to segment losses during the early stages of the TCP window growth, are prone to critical unfairness in the access to a shared medium.

4.4.4 Handover Issues

In [245], is discussed a queuing network mathematical model for the analysis of the low latency handoff schemes related to those of [246, 247, 248, 249,

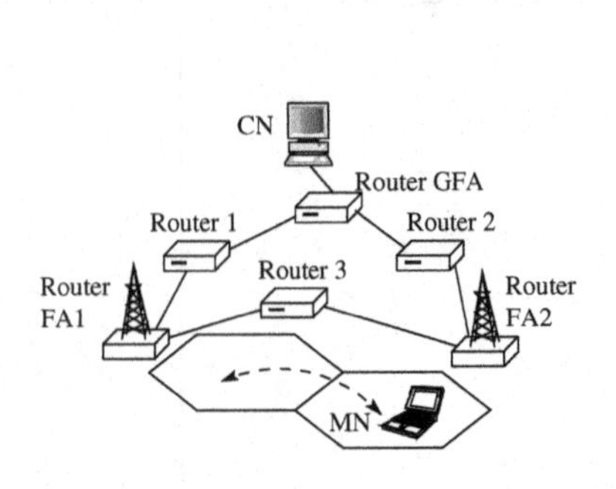

Figure 4.6: Network architecture

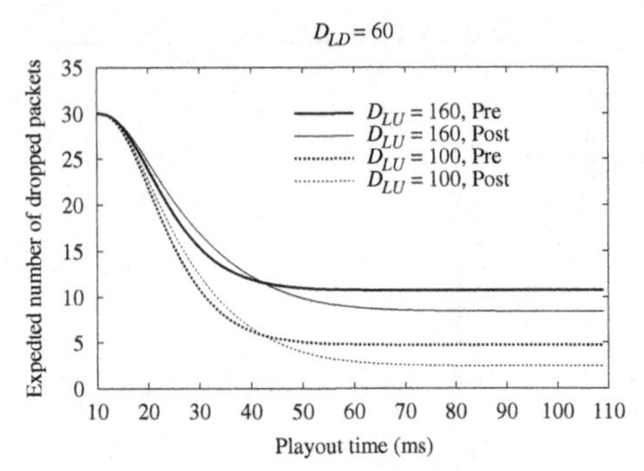

Figure 4.7: Delay distribution

250]. Consider the network architecture depicted in Figure 4.6. For computational tractability all routers are modeled as simple M/M/1 queues. The analysis focus on the Pre-Registration handoff scheme, the model corresponding to the Post-Registration scheme being presented in [249].

Consider a Mobile Node (MN) moving from the old Foreign Agent (oFA) to the new Foreign Agent (nFA) and suppose an overlapping area between the two subnetworks. Assume that the L2 handoff starts when the MN enters the overlapping area, and denote this time instant by t_0. Let the variables D_{ST}, D_{LD}, and D_{LU} define the time needed, since t_0, to generate the layer 2 triggers L2-ST (*Source Trigger*), L2-LD (*Link Down*), and L2-LU (*Link Up*), respectively. These values are constant positive values such that $D_{ST} < D_{LD} < D_{LU}$. Furthermore, the response time of any router A is denoted by R_A. The time instant when the Gateway Foreign Agent (GFA) starts forwarding to the nFA, instead of to the oFA, the packets destined to the MN is denoted by t_1 and given by $t_1 = D_{ST} + R_{oFA} + R_3 + R_{nFA} + R_2 + R_{GFA} +$ fixed delays (see Figure 4.6). This expression is a sum of exponentially distributed variables and constants. Packets will be routed via the oFA or via the nFA according to whether they arrive at the GFA before or after t_1.

As an example, Figure 4.7 shows the expected number of CBR packets that are dropped due to expiration of the playout time or the absence of a buffer, as function of the playout time. Both the results for Pre-Registration and Post-Registration are shown, for two different values of the time between the LD and the LU triggers. The timing of the LD trigger is set to 60 ms. Note that these curves tend to the expected number of lost packets due to the absence of buffer capacity when the playout time tends to infinity. It can be seen that Pre-Registration implies more losses than Post-Registration, while the average delay for packets that are not lost is slightly larger for the Post-Registration scheme. The latter follows from the fact that packets using the bi-directional edge tunnels have a longer delay. More packets are lost when the time between the LD and the LU trigger increases, so more buffer capacity would be needed to avoid losses. In the paper, issues related with handoff implementations over IEEE 802.11 are also discussed.

At present, new issues of high-speed network research emerge from the convergence of wireless and wired next generation networks and the planned deployment of new integrated multi-media and Web services. In [182], is sketched a service architecture where adaptive applications are running above a distributed service management and QoS-control layer. These layers operate, in turn, on top of the transport and network layers with their corresponding functionality of resource allocation and control, and of mobility management. In the paper is described a LINUX implementation of the proposed concept in an IEEE802.11b compatible wireless LAN with Mobile IP (MIP) as network layer using a unified programming model, and is investigated the efficiency of the data transport by the TCP and UDP protocols. By means of appropriate measurement tools is illustrated the impact of the changing transmission quality, the resulting error recovery of the wireless data link layer, and the roaming between different basic service areas on the dynamics of TCP flow control and on the resource-reservation process taking into account realistic emission patterns of the IP frames. Moreover, the most effective and sensitive control information of the adaptation process is identified.

As a summary, the study identifies some important issues regarding the interworking between resource reservation, QoS-control, mobility-, and security-management that require an improved, more efficient solution in the near future. From experiments it is concluded that an interworking between the TCP flow-control and the behavior of the wireless Data Link Control (DLC) layer should be organized to improve the throughput and delay characteristics of the data flows while moving and roaming, e.g., by exchanging management and control information that optimizes an Explicit Congestion Notification (ECN) based TCP flow control in a way similar to the Available Bit Rate (ABR) scheme in Asynchronous Transfer Mode (ATM). The use of improved TCP variants, e.g., those that freeze and recover the congestion-window state after a handoff, that use TCP control block statistics, that mimic a lossless Protocol Data Unit (PDU) transfer at the data link layer like snoop, or that distinguish different sources of packet loss, may be another alternative.

Despite the technical shortcomings of the implementation, it is further demonstrated that micro-mobility is not supported in an adequate manner by the current functionality of MIP. Cellular IP, route and address caching techniques, improved movement detection by eager or hinted cell switching, and fast or two-phase handover schemes may partially resolve the observed difficulties. One of the most important issues is the lack of a co-ordinated interworking between MIP, the resource management, and the handover management. This problem has to be solved in an efficient way, otherwise the performance of the handoff process will deteriorate considerably and TCP flow-control may time out, causing massive performance degradation.

In conclusion, the realization of concept and the measurements clearly reveal the potential and drawbacks of the mobile-aware approach, and provide new insight on the relationship of the signal-to-interference ratio, the delay-loss characteristic of flows, and the TCP behavior, as well as the adaptation processes. The results may be used to develop improved handoff and TCP flow-control mechanisms, together with improved mobile-aware applications, for an efficient transport of real-time multi-media and interactive Web services in next generation wireless networks.

The study in [251] refers to a WLAN network based on ATM technology. The problem of extending the connection-oriented ATM technique to mobile environments is that the users may change access point to the network (handover) during an on-going connection. As cellular networks are using smaller cells to increase frequency reusability and system capacity and to decrease used powers, the rate of handovers increases and can cause the overload of the network management if all the handovers are followed by a connection set-up procedure. The above problem could be solved with more efficient exploitation of existing resources implemented with the use of suitable handover procedures. Solving the problem by reserving channels also in the neighboring cells leads to inefficient exploitation of resources, since the mobile user entering only one of the neighboring cells makes the reservation of free capacity in the other cells unnecessary. Nevertheless, the QoS has to be guaranteed for the connection in case of handover, which otherwise cannot be ensured if the mobile user enters an overloaded cell. In addition to these problems, mobile systems with non-voice type transmission also require considerable and variable bandwidth. There are several approaches to fight this problem, such as use of the virtual connection tree, shadow cluster, umbrella cell, location prediction, and handover supporting routing algorithm. The radio link related part of the handover is often in focus of research activities. Several publications address this problem, and many solutions have been published based on the idea of channel assignment and bandwidth reservation to support handover. The novelty of the scheme proposed in [251] lies in the fact that it focus not only on the handover problem at the radio interface, but also involves the wired network and extends the routing with a sophisticated new Location Prediction algorithm to speed up the handover process. A short survey is given about the

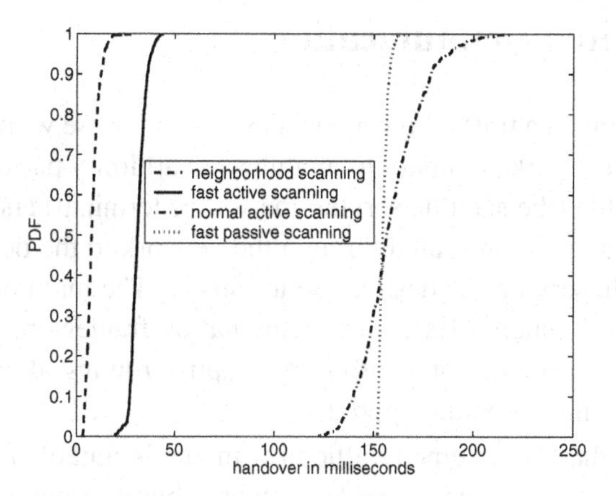

Figure 4.8: Cumulative probability distribution function of handover times using DCF

above mentioned solutions for the support of efficient handover, and a comparative study is made that shows that they can fulfil the QoS requirements. A new handover supporting routing method is also presented in the paper, and its performance investigated by means of computer simulation.

In [252], a comparison study of different Wireless LAN handover mechanisms and their ability to support QoS traffic is made. The handover mechanisms and additional protocols are implemented in a simulation environment towards the analysis of their ability to support a specific QoS level. Only two of the three parts of the handover are analyzed, since a form of pre-authentication for the simulations is assumed. The study concentrates on the analysis of the different scanning mechanisms, since the scanning process is shown to dominate the handover time. The handover performance is analyzed in relation to three different traffic types, no background traffic, voice traffic, and a traffic mix with FTP data sources. It is shown that a 50 ms Beacon inter-arrival time does not decrease the maximum throughput, while highly improving the handover performance for the passive scanning mechanisms. However, the normal passive scanning still does not suffice for satisfying the QoS requirements. Figure 4.8 shows the cumulative probability distribution function of the complete handover times when using the Distributed Coordination Function (DCF).

It can be concluded from this figure that neighborhood scanning provides the fastest handover, but this scanning mechanism is not yet included in the IEEE 802.11 standard and fast active scanning is completely sufficient for providing QoS in public hot spots. The studies prove that QoS support in Wireless LAN environments is possible even if the station has to perform a handover. Further studies have to take a closer look at prioritizing multimedia traffic and analyze the system performance in case of overlapping and co-located cells.

4.5 Satellite Communication

In [210] is described a traffic flow model developed for use with an Inter Satellite Link (ISL) network simulator to modulate an arbitrary packet generator on the origin satellite (the satellite serving the source terminal) and thus generate data packets that are then routed through the network to the destination satellite (the satellite serving the destination terminal). The main objective of this work is the development of a complex simulation framework for testing and analyzing the performance of an arbitrary adaptive routing algorithm, hence it was built following a modular approach.

It is shown that the assumed traffic flow model is suitable for performance evaluation of global satellite networks with an arbitrary satellite constellation. However, because of specific implementation issues, the study focused on a Low Earth Orbit (LEO) satellite system based on an inclined (delta pattern) constellation of 63 satellites in 7 orbital planes. In this configuration, the orbital planes have an inclination angle of 48° with respect to the equatorial plane, and satellites orbit the Earth at 1400 km with an orbit period of 114 minutes. The ISL network assumes permanent topology with two intraplane and two interplane ISLs per satellite.

In [210] is addressed the issue of per-hop packet routing, a technique particularly well suited to the regular mesh topology with multiple alternative paths of connectionless ISL networks. In particular, it is proposed the use of Traffic Class Dependent (TCD) routing, capable to find suitable paths that can accommodate different types of services using different optimization criteria. However, similarly to conventional single-service routing procedures deployed in packet networks, TCD routing still does not guarantee the provision of any minimum requirements as in QoS routing. Three different representative traffic classes with diverse requirements have been introduced to evaluate the performance of the TCD routing procedure in the ISL network, each being routed according to its particular optimization criteria.

The performance of the proposed TCD routing procedure in different traffic load scenarios is evaluated using a simulation model of the ISL network and compared with the performance of a simple single-service routing, which makes no distinction between packets belonging to different traffic classes. Also, results obtained with empirical traffic are compared with those obtained using equivalent Poisson traffic. In both cases, homogeneous and non-homogeneous traffic flows between satellites are considered. Simulation results are presented as a trade-off of two performance measures, average relative packet delay deviation and average normalized data throughput.

The routing strategies and scheduling policies for the ISL network of a packet switched non-geostationary satellite system with voice, video and best effort traffic classes, are analyzed in [83]. The case studies cover different combinations in routing strategies (Single Service or TCD routing) and scheduling policies (FIFO, Priority, Fair Queuing). To evaluate the performance of different combinations of scheduling and routing procedures, a specifically developed model in the OPNET Modeler simulation tool was used to model the ISL network. This model was built on the packet level. The analysis was done in terms of average packet delay and average data throughput for different traffic scenarios. The results imply that the introduction of TCD routing with appropriate scheduling policy significantly improves the network performance for all traffic classes as compared to a single service routing, while there is a small impact of different fair scheduling mechanisms on network performance in the case of TCD routing.

4.6 Bluetooth Technology

The number of devices making use of Bluetooth technology has rapidly increased thanks to the amount of implementations in different types of small devices. Especially in a heavy industrial environment, typically with lots of signal disturbances, Bluetooth works perfectly thanks to the Frequency Hopping Spread Spectrum (FHSS) technique. The FHSS technique avoids disturbance by using pseudo-random jumps between frequencies over the spectrum of 2.402–2.480GHz. Another advantage is the resistance to undesirable propagation conditions. FHSS together with error correction could also allow for more users than the number of available frequencies, even though the probability of a collision will increase. Still, the error correction coding could recover the transmitted packet.

A good way to understand the concept of FHSS and its behavior is to undertake the task of creating a simulation model and to understand the inner nature of the technique. Before any study of data from a simulation model can be performed, one has to make sure that it is accurate and that trustworthiness is established. Thus, one issue of concern arises when evaluating performance data derived from a simulation model: Is the simulation model correct? Can the results from the simulation be trusted? Are they fairly accurate? These questions are important because if the simulation model is wrong, all simulations derived from it may not be trustworthy. During a verification and validation of a simulation program, it is important to use different techniques to establish a credible model. When more techniques are used and results match, the simulation model in question is more credible. These issues are examined systematically in [253].

Using the validated and verified Bluetooth (IEEE 802.15.1) simulation model, a study on how to reduce the time-to-connect is carried out in [254]. A wireless ad hoc network for Bluetooth is formed without the help of any fixed infrastructure. For Bluetooth this is called a Piconet, and can involve several devices. Regarding Bluetooth, the so-called state machine, which is one of a few parameters inside the selection box, is responsible for the Frequency Hopping (FH) calculation. In this state machine, a particular sub-state, Standby-Random Backoff (RB), is heavily involved in the time-to-connect between two Bluetooth devices.

Before one can use an ad hoc network with Bluetooth devices, one or several connections must be established. This is done by using an inquiry, page, and connection procedure. A pre-research has been executed regarding the ratio each procedure occupies. The result indicates a share of 56 % for the inquiry procedure, 43 % for the page procedure, and 1 % for the connection procedure. This indicates that the efforts of reducing the time-to-connect should be concentrated onto the inquiry procedure. According to the Bluetooth specification, a maximum value of 10.24 s for the inquiry procedure may be necessary to find a Bluetooth device within range. This also motivates the needs of an improved time-to-connect for Bluetooth in the inquiry procedure.

The sub-state Standby-RB in the inquiry procedure is investigated thoroughly. Different parameters together with the RB parameter are used to simulate its behavior. Finally the results are then used in the above-mentioned Bluetooth simulator to simulate the inquiry procedure. When the default RB boundary is used, several large inquiry time values are discovered, and with an optimized RB boundary these large values have disappeared. The results show the importance of defining proper boundary values of the RB parameter in order to yield an optimal behavior.

4.7 Mobility Issues

This section includes the issues related to user mobility that are essential for performance analysis of wireless networks and are not technology specific.

4.7.1 Mobility Models

In [218, 219] are discussed questions related to user mobility models. Mobility models are needed in the analysis of wireless mobile networks, whether the approach is numerical or analytical. The mobility of the nodes has direct consequences, e.g. to the traffic and signalling loads in different parts of the network. The mobility models are especially important in the analysis of mobile ad hoc networks, where the network connectivity depends on the current locations of the nodes.

Several mobility models have been proposed and studied in the literature. Generally the models can be classified into two groups, synthetic models and models based on actual measured traces. The so-called Random Waypoint Model (RWP), originally proposed in [255], is one of the most widely used synthetic mobility models in performance analysis of ad hoc networks.

One intrinsic property of any mobility model is the distribution of the node location (or shortly the node distribution), i.e. the distribution which describes where the nodes are located on average. The uniform distribution is commonly assumed in many performance studies, for example in studies on ad hoc network capacity (see, e.g., [256, 257]) and connectivity properties of random networks (see, e.g. [258, 259]. However, the spatial node distribution resulting from the RWP process is heavily concentrated in the middle of the movement area. Thus, knowledge of the actual node distribution is often needed in order to study the impact of mobility on the performance measure of interest. For the RWP model in the plane, the stationary node distribution has been studied, e.g. in [260, 261, 262, 263, 264].

The traditional RWP model defined by a sequence of random waypoints, P_1, P_2, ..., placed randomly using a uniform distribution in some convex domain $\mathcal{D} \subset \mathbb{R}^2$. At any point of time, the node moves at constant speed v along a line from waypoint P_i towards the next waypoint P_{i+1}. Once the node reaches waypoint P_{i+1} it takes a new heading towards the waypoint P_{i+2}, etc. Each transition from one waypoint to another is referred to as a leg.

In [218], the stationary spatial distribution of a node moving according to the RWP model in a given convex area is studied. An exact expression is found, which is in the form of a one-dimensional integral, giving the density up to a normalization constant. The normalization constant is a function of the size of the area and the mean leg length. Additionally, a modified RWP model is studied, where the waypoints are on the perimeter. The analytical results are illustrated through numerical examples on a unit disk, on a rectangular area and on a irregular pentagon. Finally, the analytical results have been utilized to derive several efficient simulation methods for generating samples from the node distribution.

In [219] and in [265], the RWP model is further extended to n-dimensional (Euclidean) space, from which only the three-dimensional case is likely to have any practical value in the context of mobility modelling. In particular, the three-dimensional RWP process can serve as an elementary model for, e.g. mobile users in an office building or a shopping center. Other possible applications include airborne and underwater objects. The main results in [219] are the analytical integral formula for the stationary distribution of a node moving according to a RWP model in n-dimensional space and a closed form polynomial solution to the integral formula in the case of unit ball in $\mathbb{R}^3$.

4.7.2 Location Management

A novel location management method, designated LTRACK, is proposed and examined in [266]. LTRACK is more efficient than the Mobile IP or the GSM location management schemes. An LTRACK network is built up from LTRACK nodes. A mobile node is connected to one of the LTRACK nodes in the network, and it can change its point of connection. Every mobile node has an entry in a home LTRACK register (HLR). The basic idea behind LTRACK is to find a compromise between the Mobile IP scheme, where the Home Agent (HA) has exact location information, and the GSM scheme, where only the Location Area (LA) and no further information is known. The HLR of LTRACK does not have exact location information, but when an incoming packet arrives, the exact location of the mobile node can be determined. For each of the mobile nodes, the HLR stores the last address from where it received a location update message. It is a *next-hop* towards the node. The mobile node is either connected to that LTRACK node, or that LTRACK node knows a *next-hop* LTRACK node towards the mobile. Once an incoming call arrives, there is a series of LTRACK nodes pointing from the HLR to the mobile node. When the mobile node moves from one LTRACK node to another, handover takes place. The LTRACK node that the mobile moves away from is called the old LTRACK node, the one it moves to is called the new LTRACK node. There are two different kinds of handover in LTRACK: normal handover and tracking handover. In a normal handover the mobile equipment updates its entry in the HLR. It sends the address of the new LTRACK node to the HLR. In case of a tracking handover the mobile sends the address of the new LTRACK node to the old LTRACK node.

4.7.3 Reliability Aspects of Mobile IP

The mobility provided by the Internet Engineering Task Force (IETF) Mobile IP cannot properly fulfil the requirements of an always-on scenario; hence micro mobility is needed to extend macro mobility. There are several micro mobility protocol recommendations introduced in the literature, most of them based on a physical or logical tree topology network. The most important and severe weakness of the tree topology is its poor reliability. In [267], the reliability of mobile IP is investigated. For this purposes a graph model of mobile

IP networks is introduced where the edges of the graph represent the links connecting the nodes. In the model only links break down, the nodes being totally reliable. The links have two states, up (working), and down (broken). In this simple model all the links have an independent, and very low probability of being in the down state. Reliability means that faults in the system do not degrade the performance of the system too much. To formalize this statement, a performance function is introduced as a reliability measure, representing the performance of the system in a state. Maximum performance is the value of the performance function in the faultless state of the system. If the performance of the system in a given state is divided by the maximum performance, we get the relative performance. The reliability measure is the following: the performance of the network in a given state is the number of base stations that can reach the backbone. For general tree topology micro mobility networks, a recursive algorithm is proposed for the calculation of the exact distribution function of the relative performance. This algorithm works for any kind of tree topology micro mobility network. Using the same reliability measure, the algorithm can be extended to other network topologies.

4.8 Data Transmission over Wireless Links

To cope with relatively high bit error rate in wireless links, special error handling mechanisms like error correction/detection codes and retransmission policies are necessary.

In [268] is studied how the link layer error handling design in wireless links affects channel energy efficiency and TCP throughput. The analysis is based on the definition of Markovian models that reproduce the radio channel dynamics, the link layer hybrid Forward Error Correction (FEC)/ARQ protocol and the TCP behavior. The aim is to find the values of the link layer parameters that optimize both the channel efficiency, measured in terms of energy consumption, and the TCP throughput. As main results, it is shown that there are no practically useful alternatives to the adoption of a fully persistent ARQ protocol on the wireless link to avoid energy waste. Furthermore, it is verified that, when a fully persistent ARQ is used, the link layer parameters that optimize the channel efficiency are optimal even from the point of view of TCP performance.

The derivation of the distribution of error events in a convolutional code at the output of a Viterbi decoder is provided in [269]. Even if the mean error behavior of convolutional codes is a well known topic, a precise analysis of the error events at the output of Viterbi decoder is still lacking. The classical approach to the derivation of the error events distribution after decoding of a convolutional code is by using the weight distribution of the code. However, this approach is specific to each code, and the complexity involved can be quickly tremendous when the memory of the code goes larger. In the paper, the error exponent theory is applied to derive bounds on the distribution of error events. These bounds are shown to be sufficiently tight to be used in practice.

Chapter 5
Optical Networks

Ezhan Karasan
Bilkent University, Turkey

Nail Akar
Bilkent University, Turkey

Contributors:

Dieter Fiems (Ghent University, Belgium), Christoph Gauger (University of Stuttgart, Germany), Guoqiang Hu (University of Stuttgart, Germany), Esa Hyytiä (Helsinki University of Technology, Finland), Veronique Inghelbrecht (Ghent University, Belgium), Konrad Laevens (Ghent University, Belgium)

5.1 Introduction

The optical network design problem explores the optimal trade-off between cost and performance considering the network architecture, size and topology, and the amount and dynamics of the traffic demands to be transported over the network. The statistical or deterministic multiplexing technique that will be used in the network and the determination of the appropriate switching granularity, considering the dynamics of the traffic and the cost of different switching architectures with varying granularities, are some of the key aspects of the optical network design problem. The architecture of the optical network can be determined using several approaches, each providing different levels of flexibility and functionality with various degrees of complexity and cost.

In fiber switching, the traffic carried over a fiber is switched as a whole using electro-mechanical switching technologies, e.g., Micro-Electro-Mechanical Systems (MEMS). The bandwidth granularity in fiber switching is very coarse, since a single fiber may carry up to several Terabit per second. This solution is fairly simple from the implementation point of view and inexpensive from the switching cost aspect, but offers a very limited flexibility. On the other hand, in wavelength switching, the whole traffic transported over a wavelength in a fiber is switched together. The switching granularity with wavelength switching may be up to two orders of magnitude finer than with fiber switching, providing more flexibility and better utilization, but at the cost of higher complexity and consequently switching cost. On the other hand, the wave-

J. Brazio et al. (eds.), Analysis and Design of Advanced Multiservice Networks Supporting Mobility, Multimedia, and Internetworking, 149–166.

length continuity constraint, which requires end-to-end availability of the same wavelength in all links of an end-to-end path, limits the achievable utilization for wavelength switching networks. The utilization performance can be enhanced further by using wavelength interchanging switches, although at a significantly higher cost. Recently, waveband switching has been proposed for reducing the switching cost and complexity. In waveband switching, a group of wavelengths, called a band, is switched together through a node. Since the number of bands is much smaller than the number of wavelengths, port counts are significantly reduced, resulting in a reduced switching cost.

Meanwhile, the bursty nature of the data traffic leads to sub-wavelength switching for better resource utilization with finer granularity, which is realized by introducing new multiplexing levels, such as Optical Burst Switching (OBS) and Optical Packet Switching (OPS), and Optical Time Division Multiplexing (OTDM). All these optical technologies are currently in the stage of research and experimentation.

Different switching granularities pose different engineering problems. Fiber and wavelength switching are similar to traditional circuit switching technologies. Optical circuits, with various granularity levels, are logically associated to connections called lightpaths, and the basic network design problem is to carry as many lightpaths as possible with a given set of resources, for the dynamic traffic model, and to determine the minimum cost for network resources necessary to accommodate a given traffic load, for the static traffic model. The process of obtaining a layout for the lightpaths is called the *Routing and Wavelength Assignment* (RWA) problem, which has been studied extensively in the literature. The introduction of wavelength conversion eliminates the wavelength dimension of the problem and introduces a degree of freedom. Because of the high cost of the wavelength converters, deploying these devices at only selected nodes (sparse conversion) or deploying a limited number of converters at all nodes (partial conversion) have been considered as cost-efficient alternatives in the literature. An interesting problem in this context is the determination of where and how many wavelength converters should be deployed in the network.

Contrary to fiber/wavelength switching, sub-wavelength switching with statistical multiplexing, such as OBS and OPS, elicits problems that are more similar to those in the legacy packet-switched networks, such as resource allocation and contention resolution. Contention resolution, congestion control, scheduling, and optical packet/burst assembly are some of the widely studied problems in OBS/OPS networks.

Another method for increasing the functionality and flexibility of the network at competitive cost levels is to make use of hybrid approaches, where different technologies and switching granularities may coexist within the same network. For example, optical and electronic buffering and/or processing might coexist in an OPS network in order to have a better cost/performance trade-off. Another example of an hybrid network is an architecture where wavelength switching and OBS/OPS coexist in the same network for dealing with different amounts/granularities of bandwidth demands between node pairs, e.g., smaller bandwidth requests are fulfilled using OBS/OPS multiplexing, whereas requests with larger bandwidth are directly handled by wavelength switching.

In conclusion, various switching techniques pose different design and analysis problems. In the rest of this chapter, we present the outcomes of the research projects conducted within COST Action 279 for optical circuit and burst/packet switching networks.

5.2 Optical Circuit Switching

Optical circuit switching is used for large-sized aggregated traffic. Once the optical circuits, i.e., lightpaths, are established, the formed logical topology, consisting of these lightpaths, does not need to be reconfigured frequently. This provides cost-efficiency in the core network because the number of add-drop points necessary for transporting the traffic is significantly reduced, since the majority of the traffic in core networks, e.g., 60–80% of the total traffic incident on a node, typically expresses through that node [270]. Optical Cross-Connects (OXC) perform the switching operation completely in the optical domain and do not require costly Optical/Electronic/Optical (O/E/O) converters (transponders) for the express traffic, resulting in lower complexity and cost. OXCs also support bit-rate, protocol and transmission format transparency, which enables transport of various data traffic, e.g., Gigabit Ethernet, ATM, SONET, IP etc., on different channels. Transparency also provides more flexibility so that the system can be connected directly to any signal format without extra equipment. However, OXCs cannot perform grooming of low-rate traffic streams, and they may have problems with transmission impairments, since all-optical 3R regeneration (i.e., amplification, reshaping, and re-timing) still remains at a research stage. Another potential problem of OXCs is the vendor interoperability issue, since standardization of WDM interfaces is not complete yet.

Because of the limitations of building national-scale multi-vendor all-optical networks and the high cost of opaque networks, translucent networks are also considered for core networks [271]. There are two proposed architectures for translucent networks: (i) a network based on sparse placement of opaque nodes where regeneration and wavelength conversion are possible [272, 273], and (ii) a network composed of transparent islands, i.e., all-optical subnetworks, interconnected to each other via transponders [274].

OXC architectures that have been proposed in the literature can be classified with respect to the level of switching. There are three types of OXCs with varying switching granularities:

- Wavelength selective cross-connects (WSXC) and wavelength interchanging cross-connects (WIXC), capable of switching at the wavelength channel level

- Waveband switching cross-connects (WBXC), with the capability of switching groups of adjacent wavelengths combined into wavelength bands

- Fiber switching cross-connect (FXC), capable of switching all the wavelengths in an entire fiber.

Whereas WIXC has the highest flexibility among these four switches, it has the highest cost due to high-priced wavelength converters. On the other hand, FXC, which is basically an automated fiber patch panel, has the least complexity, flexibility and cost. FXCs are mainly considered for being used in protection switching against fiber cuts. The cost advantages of fiber level protection are evaluated with respect to the wavelength level protection by eliminating costly transponders at the expense of additional fibers needed due to coarse switching granularity [275]. WSXCs are significantly less expensive than WIXCs, but wavelength conflicts may occur due to lack of wavelength converters, which should be solved using a proper RWA algorithm [276, 277, 278, 279].

While a single fiber has over a Terabit per second bandwidth and a wavelength channel has over a Gigabit per second transmission speed, the core network may still be required to support connections at rates that are lower than the full wavelength capacity. The capacity requirement of these traffic connections can vary in the range from STS-1 (51.84 Mbit/sec), STS-3 (155.52 Mbit/sec), or STS-12 (622.08 Mbit/sec) up to full wavelength capacity, STS-48 (2488.32 Mbit/sec) or STS-192 (9953.28 Mbit/sec), for backbone applications. In order to reduce the network cost, efficient grooming of the low-speed connections into wavelength channels is an important traffic engineering problem [280, 281].

In order to combine the advantages and flexibility of the switching systems with different granularities, hierarchical switching node architectures handling multi-granular traffic are also considered [282, 283, 284, 285]. A three-layer multi-granular OXC (MG-OXC) consisting of an FXC, an WBXC, and an

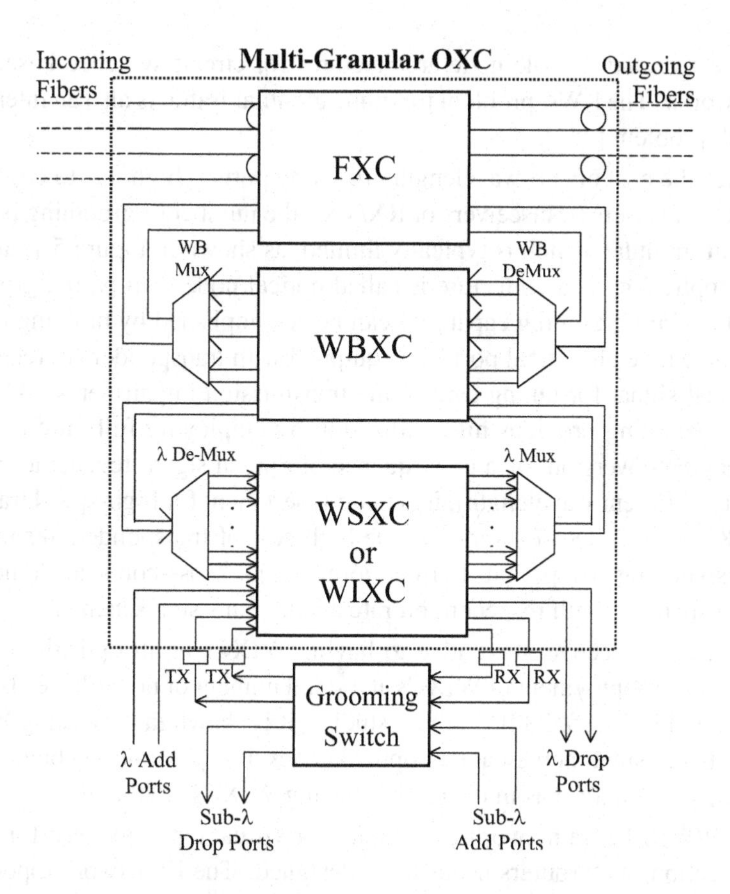

Figure 5.1: Optical node architecture with a multi-granular OXC and a grooming switch

WSXC/WIXC together with an electronics based grooming switch is shown in Figure 5.1. In this architecture, an incoming fiber is demultiplexed into its wavebands (wavelengths) using waveband (wavelength) demultiplexers if and only if the wavebands (wavelengths) in the fiber are switched to different output ports. If all the wavelengths in a fiber express through the node over the same fiber, the fiber is not demultiplexed and it is switched by the FXC. The wavelengths are switched using an WSXC or WIXC. Some wavelengths are added/dropped at the node and some wavelengths are fed into the electronic-switching capable grooming switch for sub-wavelength grooming and adding/dropping sub-wavelength traffic.

It was shown that traffic grouping with end-to-end and intermediate bundling of the traffic from the same source to different destinations, or from different sources to the same destination on common sub-paths, reduces the size of switching systems in networks [282]. MG-OXCs are promising for building cost-effective and flexible node architectures for circuit-switched based optical networks. The RWA problem for multi-granular traffic is also an interesting research problem [286].

Since the cost of sub-wavelength grooming ports is high due to costly O/E and E/O converters (transceivers or RX/TX), the number of grooming ports on the multi-granular switch is typically limited, as shown in Figure 5.1, and this type of optical node architecture is called optical node with partial grooming capability. Full grooming capability can be accomplished by utilizing opaque switches, where all optical ports are equipped with transponders to regenerate the optical signal for taking care of the transmission impairments. Although opaque switching provides high utilization by employing full grooming and wavelength conversion (as a consequence of optical signal regeneration, possibly at a different wavelength), it is not cost-efficient for high-speed transport (OC-48, OC-192, OC-768), due to the high cost of transponders resulting in higher switching cost per bit. Furthermore, opaque cross-connects do not support transparency with respect to bit rate and transmission scheme.

As a cost-effective alternative to having WIXCs at all optical switching nodes, sparse deployment of WIXCs at a small number of nodes have also been considered [287, 288, 289]. In these studies, it is shown that by using WIXCs at a relatively small percentage of optical nodes, it is possible to obtain almost the same performance as in the case of having WIXCs at all nodes.

In IP/WDM core networks, the logical or virtual topology used for the interconnection of IP routers needs to be designed. The IP network topology is computed based on the traffic measurements and predictions. Once the logical topology is selected, the routing of the lightpaths that correspond to logical IP links is performed over the optical network. The IP topology may be reconfigured with respect to traffic pattern changes, or the lightpaths may be rerouted in response to physical layer failures. Another issue related to topology reconfiguration is the migration problem, dealing with the transformation from an old logical topology into a new virtual topology. The multi-layer traffic engineering problem studies the design of the logical topology considering the most significant attributes of the physical optical layer [290, 291].

One possible approach for interconnecting IP and optical layers is the full separation between the two layers, which is known as the overlay model. In the overlay model, IP and optical layers have separate control planes and the two layers interact just to exchange data. Although the overlay model results in simpler control plane architecture, it also leads to the replication of functions and suboptimum network-wide solutions, since interactions between the two layers are not considered. The alternative is an integrated approach, called the peer model. In the peer model, IP and optical layers interact and co-operate in order to optimize the overall performance.

In the optical layer routing problem, optical physical layer constraints such as the Optical Signal-to-Noise-Ratio (OSNR), wavelength dispersion, fiber type, switching fabric type, and regenerator locations are considered for engineering of the optical layer. There is also a need for defining new routing parameters and constraints that best capture and reflect the distinct characteristics of optical networking. Physical layer considerations also bring several implications for the control plane.

Below, we briefly summarize the contributions of COST Action 279 in circuit-switched based optical networks.

5.2.1 Subnetwork Partitioning and Section Restoration in Translucent Optical Networks

The problem of designing translucent optical networks composed of restorable, transparent subnetworks interconnected via transponders is addressed in [274]. In this study, the problem of designing restorable subnetworks in translucent networks is formulated as an Integer Linear Programming (ILP) problem, where the subnetworks are determined such that each one satisfies size and connectivity constraints. Section restoration, which is a sub-path protection scheme [292], is proposed for translucent networks where failed connections are rerouted inside the subnetwork containing the failed link. Section restoration provides smaller restoration times compared to the end-to-end restoration. The network design problem of determining working and restoration capacities with section restoration is formulated as an ILP problem. Numerical results show that section restoration generate fiber costs that are close to those with the path restoration technique. It is also shown that the number of transponders with the translucent optical network is substantially reduced compared to the opaque network architecture.

5.2.2 Placement of Wavelength Converting Nodes in Optical Networks with Sparse Conversion

Scalable solution techniques for designing optical networks with sparse wavelength conversion are studied in [288] and [289]. The overall approach consists of two phases. In the first phase, the routing subproblem which minimizes the total fiber cost is solved, assuming that all nodes have wavelength conversion capabilities. The computational time for this stage is substantially reduced by employing demand aggregation and cut-set inequalities. In the second stage, a minimum number of nodes is equipped with wavelength converters such that the total fiber cost in the network is the same as the optimum solution given as the output of the first stage. To this end, two different design approaches for the second stage are studied: ILP [288] and Tabu Search based techniques [289]. The effects of topology, number of wavelengths, and number of demands are studied, and the locations of wavelength converting nodes are investigated. It is shown for a 32-node mesh topology using different traffic matrices that placing WIXCs in a maximum of about 10% of all nodes is sufficient to achieve the same fiber cost as having full conversion, i.e., all nodes containing WIXCs. The performance of the Tabu Search based algorithm is compared with that of the ILP based algorithm. It is shown that the former achieves the optimum solution in more than 70% of all cases tested, and places on the average less than 10% more WIXCs compared to the optimum solutions. It is also observed that the locations of the WIXCs placed with different traffic matrices in the optimum solutions are confined to a small number of nodes, and there is a correlation between the locations of these nodes and the amount of express traffic through the node.

5.2.3 Logical Topology Design in Optical Mesh Networks

The mesh logical topology design problem in overlay optical mesh networks is studied in [293]. The inputs to the design problem are a set of nodes, corresponding to IP routers, an associated traffic matrix among these nodes, and the number of lightpaths that can be established at each node. The output of the design procedure is a logical topology determining which node pairs are to be connected via lightpaths. Traffic between node pairs that are not directly connected by a lightpath must use a sequence of lightpaths through intermediate nodes. Such multi-hopping has a detrimental cost impact on the performance of the IP network because it increases the number of packets handled by routers. Therefore, the mesh topology design problem in overlay networks is formulated in terms of minimizing the overall amount of multi-hopped traffic. Obtaining the optimum solution for the ILP formulation of the problem is outside the capabilities of off-the-shelf ILP solvers for even small problem sizes, e.g., networks with 10 nodes. In this study, Tabu Search based heuristics and lower bounding techniques are used to bracket the optimum value when no exact optimum can be obtained for networks with 20 and 30 nodes. It is shown that the heuristic algorithm generates results that are within 2–3% of the lower bound for the optimum solution. This study is also extended to the minimization of the number of lightpaths in the logical topology, subject to nodal degree and delay constraints. Similar to the approach used earlier, Tabu Search based heuristics and lower bounding techniques are used to bracket the optimum solution.

5.3 Optical Burst/ Packet Switching

All-optical packet switching is viewed as the long term candidate for deployment of optical networks carrying IP based bursty traffic. OPS enables access to the network resources at the finest bandwidth granularity, and improves the bandwidth utilization of underutilized high-capacity optical links by introducing statistical multiplexing in the optical layer. However, its implementation poses a number of technological problems. Therefore, the network architecture and switching systems for OPS networks should be designed considering the limitations imposed by existing or forthcoming optical devices.

From the networking point of view, the most widely studied problem in the literature is the contention that occurs, due to the unavailability of optical Random Access Memory (RAM), when multiple optical packets should be switched to the same output channel. A typical assumption made in the analysis of OPS networks is the slotted-time synchronous operation. The synchronous architecture provides maximum control and better performance for the efficiency of the statistical multiplexing. However, synchronous operation requires a more complicated interface with legacy protocols such as IP, e.g., the variable length IP packets cannot simply be encapsulated into fixed length optical packets. Therefore, recent research has shifted towards the case of variable length optical packets, addressing the best possible trade-off between implementation complexity, performance of the contention resolution techniques, and ease of interworking with legacy networks.

Since optical RAMs are not currently available, Fiber Delay Lines (FDL) have been considered for resolving conflicts between optical packets in the time domain. However, packet switching architectures using long FDLs are not feasible, since such FDLs require large spaces that increase the cost of optical switches. Another method for contention resolution exploits the wavelength domain, since each fiber carries multiple wavelengths that can be used for forwarding the optical packets to the next switching node. The traffic to be transmitted over a fiber can be distributed over different wavelengths and the contention between optical packets can be resolved using wavelength converters.

The traditional approach for the control layer in OPS networks assumes that the OPS network is able to switch packet payloads in the optical domain, but requiring electronic processing of the packet headers. On the other hand, there are recent proposals for OPS network architectures with all-optical header processing, where an optical signal, e.g., a subcarrier modulated signal, is transmitted in front of the optical payload in order to control and configure the switching node before the optical payload is switched. However, these technologies are still at a very early stage.

A major problem in the design of next generation IP networks is the mismatch between extremely high optical transmission rates and the relatively slower switching speeds of optical switching nodes. OBS [294] is a sub-wavelength transfer mode that is an intermediate point between optical circuit switching and optical packet switching. OBS separates the data and control planes in the optical and electrical domain in order to eliminate the technological problems involved in the all-optical processing of the packet header in OPS. A variable-length optical burst composed of several IP packets is aggregated at the edges of the optical network since electronic buffering is simple and inexpensive. Relatively large-sized optical bursts are used in order to avoid small size optical packets, so that the stringent requirements for switching speed and synchronization in the optical domain can be avoided.

In OBS, a control packet is transmitted, before the optical burst is released, on a signalling channel separate from the data channels, i.e., out-of-band signaling is used. There is an offset time between the control packet and the burst so that the OBS switches along the path from the source to the destination can be configured beforehand for the incoming burst. The control packet is processed in the electronic domain, thus avoiding all-optical header processing. The offset time between the control packet and the optical burst should be large enough for the configuration of the switches to be completed before the burst arrives. OBS is essentially based on a one-way unconfirmed reservation scheme, and may lead to contention occurring at the OBS switches, possibly resulting in partially or fully lost bursts. Contention resolution schemes involve wavelength conversion (wavelength domain), deflection routing (space domain), and FDL buffering (time domain) [295].

Reservation of an output wavelength in OBS networks is usually performed using Immediate Reservations, e.g., Just-In-Time (JIT) scheduling or Delayed Reservations, e.g., Just Enough Time (JET) scheduling. In JIT, an output wavelength is reserved for the burst immediately after the arrival of the control packet. On the other hand, in JET the wavelength is reserved for only the duration of the burst at the estimated burst arrival time. Thus, whereas immediate reservation protocols only permit a single outstanding reservation for each output wavelength, delayed reservation schemes allow multiple control packets to make future reservations on a given wavelength provided that these reservations do not overlap in time. In the delayed reservation technique, the control packet should providing the arrival time and duration for burst transmission [296]. If wavelength conversion capability is available, there is a choice of wavelengths for which the reservation can be made.

Several scheduling strategies can be devised for delayed reservation schemes.

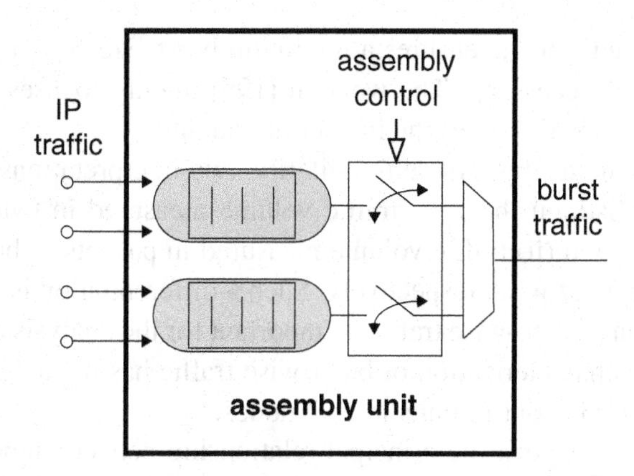

Figure 5.2: Block diagram of an assembly unit

The simple Latest Available Unscheduled Channel (LAUC) selects the wavelength with minimum residual life from the set of busy wavelengths that can be used to route the burst [297]. In order to enhance the throughput performance, the gaps or voids between bursts can be used to accommodate the incoming burst, leading to the LAUC with Void-Filling (LAUC-VF) algorithm. OBS also allows for Quality of Service (QoS) differentiation, by either using different offset times per priority class, or using preemption at the OBS switch [298]. A tutorial on OBS is provided in [299].

5.3.1 Analysis of Optical Burst Assembly

Classification and proper assembly of small IP packets to larger optical bursts at edge nodes are essential for the performance of burst reservation, transmission and electronic control in core nodes. In the assembly unit, shown in Figure 5.2, incoming IP packets are classified based on the egress node and QoS class and stored in the assembly queues accordingly. In early work on burst assembly, it has been heavily discussed whether burst assembly in edge nodes of OBS networks reduces the self-similarity of the traffic [199].

For burst assembly, two assembly algorithms can be used as basic building blocks: length-based and time-based schemes [1]. In the former scheme, a burst is sent out when sufficient amount of IP packets have been collected in the assembly queue such that the size of the resulting burst exceeds a threshold of $S_{\max}$ bytes. In the latter scheme, a time-out interval $T_{\max}$ is set upon arrival of the first IP packet to an empty queue, and a burst consisting of all packets in the assembly queue is sent out as soon as the timeout occurs. A variation

[1] A hybrid scheme which is a combination of these two schemes can also be considered.

of the time-based scheme ensures a minimum burst size $S_{\min}$ by introducing padding bytes if necessary. The work in [199] mainly focuses on these two algorithms, but also discusses the impact of padding.

Regarding the analysis of self-similarity, two interpretations of the traffic processes are distinguished: (i) traffic volume measured in bytes is referred to as *byte-wise*, and (ii) traffic volume measured in packets or bursts is called *packet-wise* or *burst-wise*, respectively. Such a differentiation is sensible, e.g., characterization of byte-wise traffic is important for the analysis of burst transport, while the characterization of burst-wise traffic has impact on the performance of electronic control units in core nodes.

In analytical models, the principal relationship between input packet and output burst traffic of a burst assembler is described in [199]. In the simulation studies, synthetic traffic and a real IP trace are used as input packet traffic to the burst assembler. In both models, byte-wise and packet/burst-wise traffic are treated separately. Analysis and simulation studies yield the following consistent results concerning the degree of self-similarity of burst-assembled traffic:

- For byte-wise burst traffic, both length-based and time-based algorithms produce the same degree of self-similarity as the byte-wise packet traffic from which burst traffic is assembled. This can be seen from the overlapping of wavelet energy curves of the byte-wise traffic in Figure 5.3 and Figure 5.4 at large time scales. Padding can principally only influence byte-wise burst traffic. However, simulations show that the self-similarity does not even change for large values of padding.

- In case of the pure length-based algorithm, burst-wise traffic has the same self-similarity as byte-wise packet traffic, which is indicated by same slope of wavelet energy curves of the burst-wise traces in Figure 5.3.

- In case of the pure time-based algorithm, burst-wise traffic may have a smaller degree of self-similarity than packet-wise traffic. Figure 5.4 shows that the slope of the curve of the burst-wise trace decreases with the increasing timeout interval τ, indicating a smaller degree of self-similarity.

In summary, only in case of time-based assembly and regarding the burst departure process, self-similarity is reduced when increasing the time-out interval. In all other cases, especially regarding byte-wise traffic, assembly has no impact on the self-similarity. Thus, in general burst assembly does not reduce self-similarity.

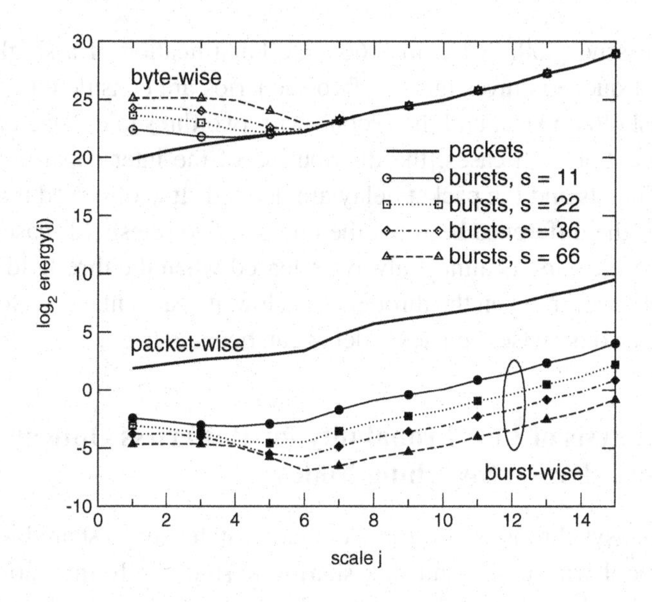

Figure 5.3: Wavelet energy curves for length-based assembly

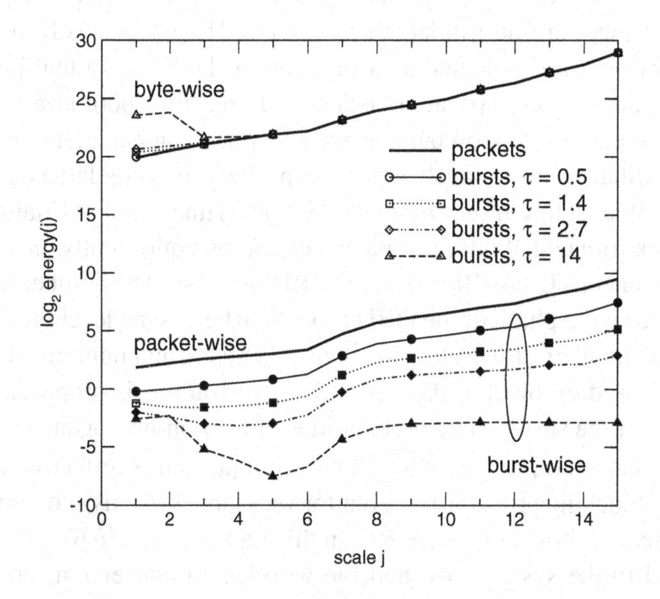

Figure 5.4: Wavelet energy curves for time-based assembly

In [166], an OBS edge router is viewed as a composition of multiple Burstification Units (BU). Each BU consists of a set of separate output queues. Studies show that the assembly strategy, which determines performance parameters such as the burst size and burst inter-departure time distributions, is a key design issue [300, 301]. In [166], the burstification of a single isolated output burst queue is investigated. Two scenarios are considered: one with a single threshold on size, and the second with two thresholds, one on size and one on age. For both cases, the distribution of the inter-departure time between two bursts and the packet delay are derived. It is observed that for high throughput, the difference between the one and two thresholds cases is rather small, since the burst is almost always released when the threshold on size is triggered. However, when the throughput is low, it is advantageous to have two thresholds, as otherwise the packet delay can become large.

5.3.2 Analysis of Loss Probability for Bufferless Optical Burst/Packet Switching Nodes

In [167], an asynchronous bufferless optical switch using a shared wavelength converter pool with per-output-link sharing is studied. In this model, an incoming optical burst, or optical packet, is blocked because either there is no available wavelength on the outgoing link, or the incoming burst requires conversion but there are no available converters. The exact steady-state blocking probabilities are evaluated as a function of the basic system parameters, e.g., mean arrival rate, arrival statistics, and converter pool size. Using the traditional model of Poisson burst arrivals, exponential burst lengths, and uniformly distributed burst wavelengths, the problem is formulated as obtaining the steady-state solution of a finite Continuous-Time Markov Chain (CTMC) with a block tridiagonal infinitesimal generator, or, equivalently, that of a finite non-homogeneous Quasi-Birth-Death (QBD) process. The number of converters in use form the phase of the QBD process, whereas the level process is dictated by the number of wavelengths in use. A stable and numerically efficient technique based on block tridiagonal LU factorizations is proposed for exact calculation of the steady-state probabilities of the non-homogeneous QBD.

Numerical examples show that this technique can be effectively used for computing blocking probabilities even for very large systems and rare blocking probabilities, as shown in Figure 5.5. In this figure, $\rho = \lambda/\mu K$ and $r = W/K$ correspond to the system load and the wavelength conversion ratio, respectively, where λ is the arrival rate of the Poisson aggregate burst arrival process, $1/\mu$ is the average duration for the exponentially distributed burst lengths, W is the size of the wavelength converter pool for each output fiber, and K is the number of wavelengths per fiber. For three values of K, $K = 50$, 100 and 200, the blocking probability P_b is plotted as a function of r and ρ in Figure 5.5 as a 3-dimensional mesh. This plot shows that, using this analysis technique, the loss probabilities can be obtained for very large systems, e.g., $K = 200$,

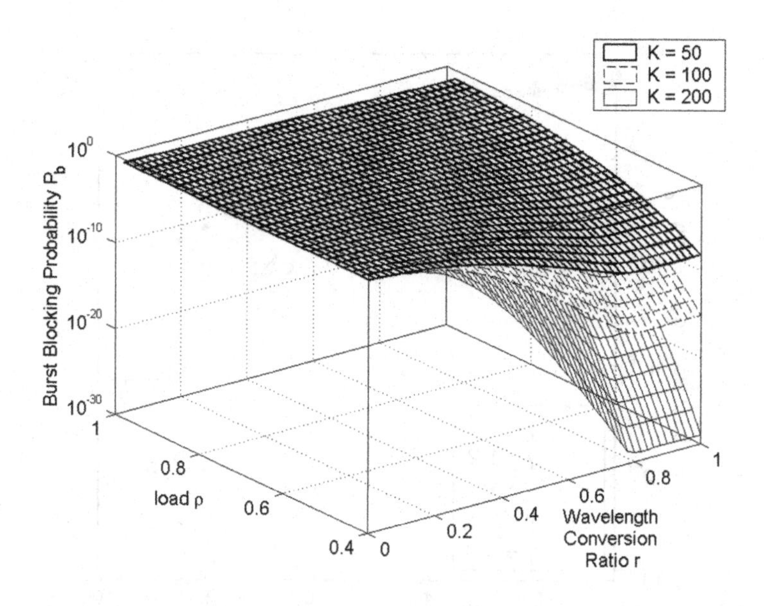

Figure 5.5: P_b as a function of the wavelength conversion ratio r and system load ρ for three different values of K

and for rare probabilities, e.g., $P_b \approx 10^{-30}$, and for a wide range of system parameters, e.g., $0.4 < \rho < 1, 0 < r < 1$. Moreover, probabilities can be obtained rather rapidly. Obtaining all three planes in Figure 5.5 takes less than an hour on a 2.4 GHz Pentium based PC.

Using the analysis technique in [167], the burst blocking probabilities are shown in Figure 5.6 for $K = 100$, $\mu = 1$, and mean burst interarrival time $= 1/60$ as a function of the Coefficient of Variation (CoV), γ, of the interarrival times. The CoV of a random variable is the standard deviation divided by the mean of that random variable and is indicative of its variability. The case of $\gamma = 1$ is for Poisson arrivals, and $\gamma < 1$ cases are obtained by using an Erlang-k distribution, for which $\gamma = 1/k$. The case of $\gamma > 1$ is obtained by using a two-phase hyper-exponential distribution with balanced means. Figure 5.6 shows that burst blocking probabilities are significantly lower for regularly spaced arrivals, i.e., small CoV. These results clearly demonstrate that the coefficient of variation of burst interarrival times is critical in burst blocking performance and therefore burst shaping at the edge of the burst/packet switching domain can be used as a proactive contention control mechanism in next-generation optical networks.

An approximate analysis of the burst loss probability in an OBS node for an arbitrary number of service classes is presented in [300]. Based on analytical and simulation results, the impact of traffic characteristics on service

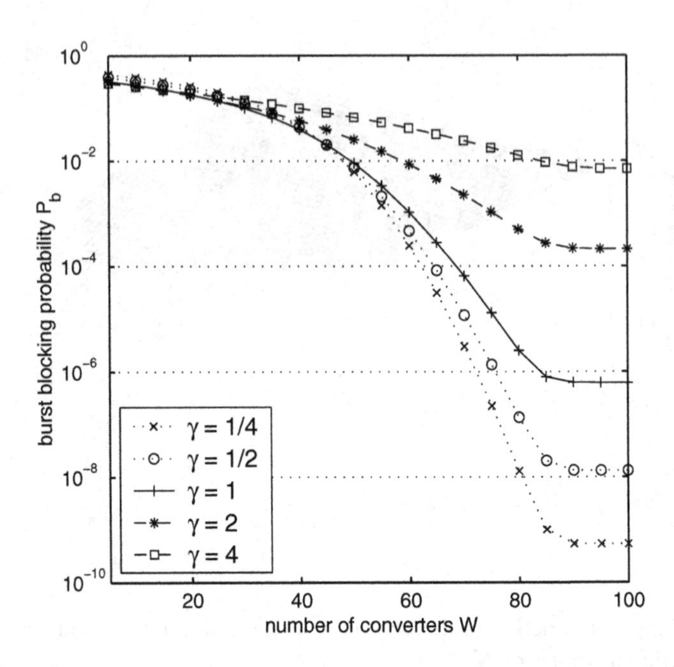

Figure 5.6: Burst blocking probabilities for $K = 100, \mu = 1$ and mean interarrival time $= 1/60$ as a function of W for different values of coefficient of variation γ

differentiation in a single node is studied. The focus lies on burst characteristics such as the mean and distribution of the inter-arrival time and burst length resulting from an assembly process at the edge of the optical network.

For a scenario with two QoS classes and a Poisson arrival process, Figure 5.7 shows the burst loss probability of the high priority class against the QoS offset normalized by the mean burst length for different low priority burst length distributions. An upper bound for the case of no isolation and a lower bound for perfect isolation are also included. First, it can be seen that the analysis matches the simulated curves quite well for all distributions. Second, the strong impact of the low priority burst length distribution, more precisely its forward recurrence time, is clearly visible. Finally, the curves show the relation of the QoS offset and the degree of QoS differentiation: for normalized offset values greater than one, the low priority burst loss probability virtually does not change. Thus, the assembly strategy has to carefully shape low priority bursts in order to efficiently operate the system.

Also, the service differentiation for various parameters in an OBS network scenario, e.g., the number of classes and the length of the basic offset, are investigated. In order to apply the presented theory to an OBS network, the multi-class analysis is applied and the traffic streams originating from differ-

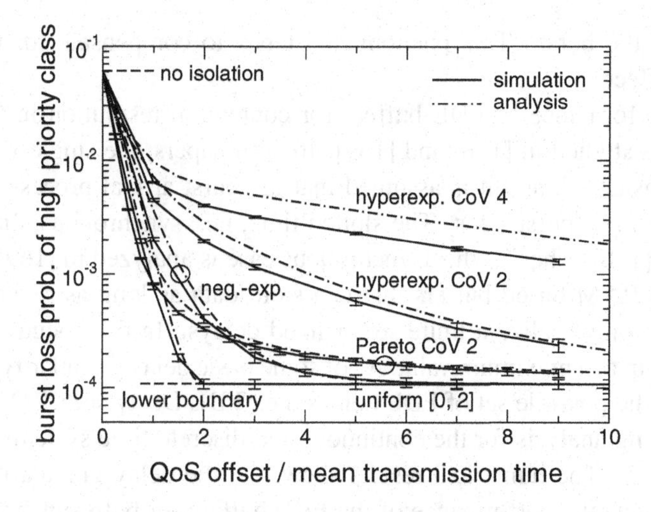

Figure 5.7: Effect of the separation/integration

ent edge nodes and belonging to different QoS classes are mapped to corresponding QoS classes. Studies show that, in a network scenario, the ratio of basic offset compensating switching and processing delay and QoS offset differentiating classes has a strong impact on the intra-class differentiation, and therefore this ratio has to be kept well below 0.1

5.3.3 Analysis of Loss Probability for Optical Burst/Packet Switching Nodes with Fiber Delay Lines

In [302], different delay line configurations and their effects on blocking probability and fairness in OBS networks are studied using a revised JET protocol with no wavelength conversion. An OBS network without FDLs cannot sustain very high traffic loads, as each link is essentially a blocking system without waiting room and with either a single server (without wavelength conversion) or multiple servers (with wavelength conversion). Thus, by adding a few FDLs into each node, the blocking probability can be remarkably reduced and the overall performance of the network significantly improved.

The main problem with protocols such as JET is that long connections have a high priority at the beginning of their journey, i.e., when bursts are close to the source node, but lose the edge as they get closer to the destination node, while in fact the opposite behavior would be desirable. This is mainly due to the reduced offset time as the burst gets closer to its destination. However, when there is congestion and a burst gets delayed by an FDL, the time between its header and the actual burst increases, which corresponds to increasing the priority of the burst. This phenomenon tends to compensate for the above negative effect.

The performance of FDL buffers for contention resolution in OBS/OPS networks is studied in [168] and [169]. In both papers, the single-wavelength case is considered, and it is assumed that the burst arrival process is memoryless with i.i.d. burst sizes. The slotted-time, i.e., synchronous, operation is studied in [168], whereas the asynchronous case is analyzed in [169].

Unlike RAM-based buffers that can store data as long as required, FDL buffers can only realize a finite set of fixed delays. In these studies, integer multiples, up to a maximum value N, of some basic delay granularity D is considered as the possible set of FDL delays, i.e., FDL Delay = $D, 2D, \ldots, ND$. The queueing analysis for the continuous and discrete-time systems are given in Chapter 2. The numerical results show that the delay granularity significantly impacts the performance of the FDL buffers for both synchronous and asynchronous cases. As a result of this delay granularity, voids are created on the outgoing channel, i.e., bursts are waiting in the FDL buffer although the channel is idle, which reduces the channel utilization. Scheduling algorithms different from First-In-First-Out (FIFO) scheduling are considered in [168] and [169] that can mitigate this effect to some degree, at the price of packet/burst reordering and increased control complexity. Loss figures are very sensitive to the exact value of D, with the optimum value of D being a function of the load and the burst size distribution. This makes a good design choice very difficult in practice.

The performance of feed-forward switches with FDL buffers is investigated in [170] and [303]. The incoming packets are routed according to the "shortest FDL first" scheduling discipline, such that an incoming packet is routed to the shortest FDL where no packets are routed to upon arrival of this packet. As this routing scheme allows output contention, a two-stage FDL buffer is investigated in [171], where the output of the first FDL buffer is routed to a second FDL buffer also using the "shortest FDL first" scheduling discipline. Numerical results show that a limited number of FDLs mitigates packet loss considerably in case of the single stage FDL structure. However, the "shortest FDL first" scheme turns out to be unsuitable for further lowering the packet loss by means of additional buffer space. A better solution is given by adding a second FDL stage as shown in [171].

Chapter 6
Peer-to-Peer Services

Kurt Tutschku
University of Würzburg, Germany

Contributors:

Andreas Binzenhöfer (University of Würzburg, Germany), Maria Luisa García Osma (Telefónica I+D, Spain), Öznur Özkasap (Koç University, Turkey), Hannu Reittu (VTT Information Technology, Finland), Kurt Tutschku (University of Würzburg, Germany).

6.1 Introduction

Peer-to-Peer (P2P) services have become very popular recently, as witnessed by the relentless spread of the Gnutella [304], KaZaa [305], and eDonkey [196] file sharing applications. P2P services have even surpassed the World Wide Web (WWW) in popularity, at least in terms of traffic volume. Backbone operators, Internet service providers, and access providers consistently report P2P-type traffic volumes exceeding 50% of the total traffic load in their networks [306, 307, 308], sometimes even reaching 80% at nonpeak times [309, 310]. These figures reveal that P2P services have evolved into one of the most popular applications in today's Internet, thus making P2P an important area of networking research.

The aim of this chapter is to summarize the research on P2P applications performed in the COST 279 community. This chapter discusses first in Section 6.2 the concept, the basic functions, and the architecture of P2P services. Section 6.3 is dedicated to the traffic patterns of popular P2P services. In particular, the characteristics of Gnutella overlays (Section 6.3.1) and of the eDonkey file sharing application in wireline and wireless networks (Section 6.3.2) are investigated. The performance of selected P2P mechanisms is investigated in Section 6.4. The efficiency of a Chord-like resource mediation algorithm is discussed in Section 6.4.1. Section 6.4.2 is devoted to the discussion of the objectives and the ways in which an Internet service provider can influence and improve the resource access in P2P file sharing applications. Section 6.4.3 describes an P2P anti-entropy algorithm for content distribution and shows its performance.

J. Brazio et al. (eds.), Analysis and Design of Advanced Multiservice Networks Supporting Mobility, Multimedia, and Internetworking, 167–177.

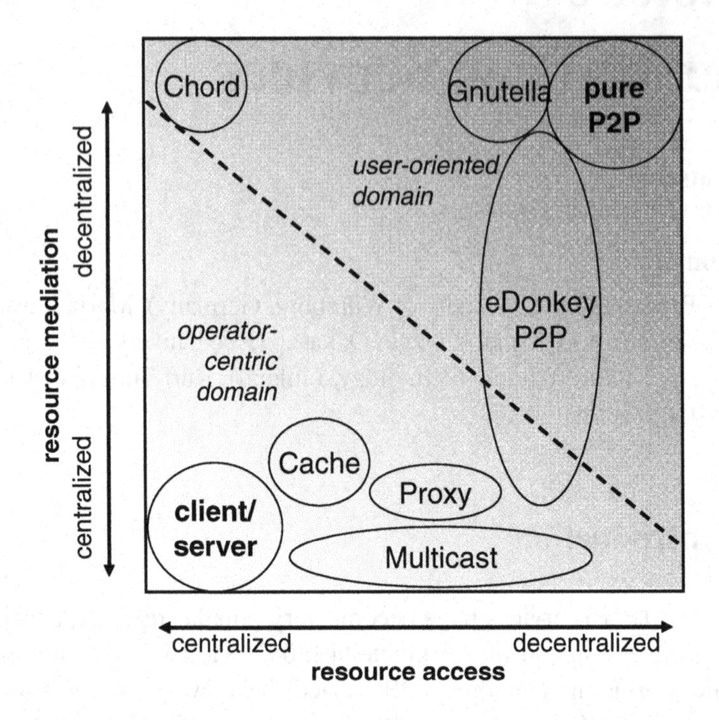

Figure 6.1: Cartography of P2P applications and content distribution architectures

6.2 The P2P Concept

P2P is a highly distributed application architecture where equal entities, denoted as *peers*, voluntarily share resources, e.g., files or CPU cycles, but also meta-data, e.g., location of files, via direct, end-to-end exchanges. The advantage of P2P services is the autonomous, load-adaptive, and resilient operation of these services. In order to share resources, P2P applications support two fundamental coordination functions: (i) *resource mediation* mechanisms, i.e., functions to search and locate resources or entities, and (ii) *resource access control* mechanisms, i.e., functions to permit, prioritize, and schedule the access to resources.

P2P architectures cover a broad range of choices. *Pure P2P* architectures, such as used in Gnutella, implement the two control mechanisms in a fully decentralized manner [311], while *Hybrid P2P* systems utilize central entities, e. g., the eDonkey index servers, that collect mediation data, such as file location information.

Figure 6.1 provides a two-dimensional cartography for comparing P2P architectures with other well-established information dissemination mechanisms. The basic P2P control functions (*resource mediation/resource access control*) form two orthogonal axes in Figure 6.1. The degree of distribution (*centralization/decentralization*) is used as the range of the axes.

In addition, the cartography visualizes the architectural choices of operators and users for providing information distribution services: the domain of *operator-centric* architectures, which aims at a strong centralization of control, and the domain of *user-centric* architectures, which typically endeavors a strongly decentralized nature of control.

The cartography reveals the focus on user-oriented services of P2P architectures, whereas the most common information dissemination mechanisms are using an infrastructure typically operated by provider.

A major reason for the success of P2P systems is their operation on the application level and the forming of *application-specific overlays.* P2P overlays work without specific network or transport support, and can be run completely at the edge of a network. While these overlays do implement a certain type of group communication structure, they do not suffer from the same deployment difficulties as multicast did in the past. However, ease in deployment comes at a cost: P2P services trade their advantages with the disadvantage of causing high data and signalling traffic volumes. The traffic patterns of P2P applications fluctuate strongly in time and space. As a result, it is anticipated that traditional network design techniques and traffic engineering procedures may no longer be applicable and new methods may be needed that maintain the autonomous and *self-organizing* characteristics of P2P while properly dimensioning the networks.

Although P2P is currently mainly related to doubtful "royalty free" access to resources, P2P generates indirectly a significant amount of revenue in networks, e.g., by the increased use of digital subscriber lines. Thus, P2P file-sharing might also become highly attractive for network operators, in particular mobile operators, since they search for new applications which do both: (i) exploit, qualitatively and quantitatively, the potential of new technologies, and (ii) motivate the user to adopt these technologies.

The application domain of P2P mechanisms in networks is not only restricted to file sharing. It may also comprise the use of self-organizing P2P mechanisms for network control. For example, *Distributed Hash Tables* (DHT) [312] can be viewed as an alternative way to locate users or resources in a Voice over IP (VoIP) architecture. The efficiency of such a mechanism was just recently demonstrated impressively by the P2P VoIP application Skype [313]. Skype replaces central index servers by a distributed, self-organizing storage mechanism in the user's end system. The use of software instead of servers might reduce the capital expenditures (CAPEX) as well as the operational ex-

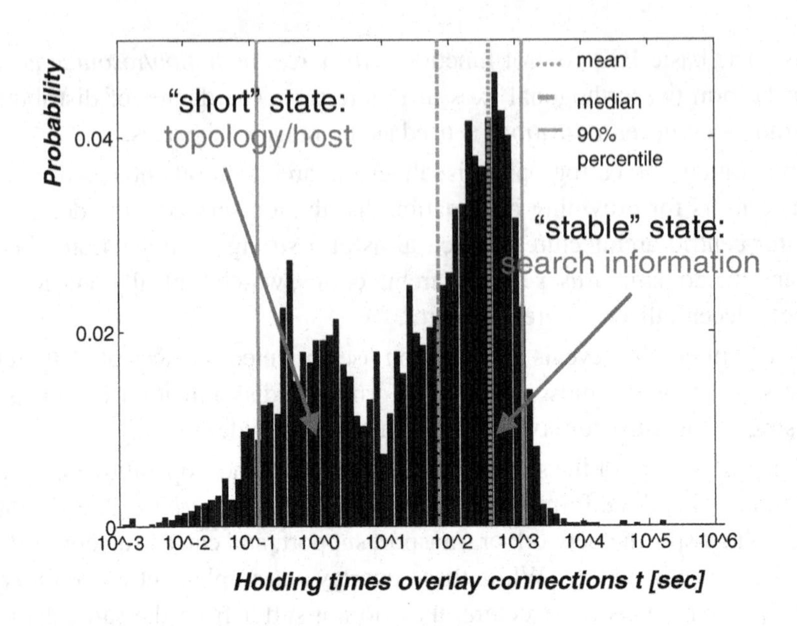

Figure 6.2: Gnutella overlay connection holding time distribution

penditures (OPEX) in networks, since fewer entities have to be operated for a new service.

6.3 P2P Traffic Characterization

6.3.1 Gnutella Traffic

In [197], a measurement study is presented on the signalling traffic in Gnutella overlay networks. Both signalling load and the scale of variability in the existence of P2P overlay connections are investigated. The study reveals that an uncontrolled Gnutella client consumes high amounts of bandwidth for its signalling traffic, reaching up in the order of tens of Megabits per second. Furthermore, the signalling traffic in Gnutella overlays varies strongly over short timescales, due to the Gnutella use of flooding protocols. The investigation of the overlay connection holding time in Gnutella showed that the distribution typically has bi-modal characteristic, cf. Figure 6.2. The modes correspond to a "short" state, where typically host information is transmitted, and to a "stable" mode, where mainly content queries are exchanged. The modes identify the time scales on which a dynamic and adaptive management of P2P overlays and P2P services is of advantage or needed.

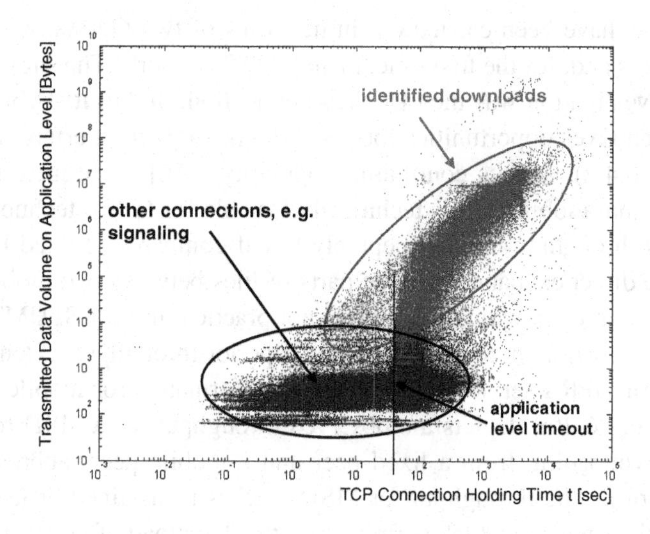

Figure 6.3: Correlation of eDonkey TCP holding time and flow size

6.3.2 eDonkey Traffic

A traffic profile for the eDonkey P2P filesharing service is presented in [196]. The eDonkey system is typically used for exchanging very large files like audio/video CDs or even DVD images, and possesses a hybrid P2P architecture with distinct servers and clients. The eDonkey system makes use of the multi source download feature, which permits the simultaneous transmission of file chunks to a downloading peer. The traffic profile shows that signalling and download have significantly different characteristics. Figure 6.3 depicts a scatter plot describing graphically the correlation between the Transmission Control Protocol (TCP) holding time and the size of eDonkey flows. Each dot in the scatter plot represents an observed eDonkey flow. The brighter dots are identified download flows, the dark dots represent non-download connections. The scatter plot shows that almost all identified download flows are within the same region. In turn, the non-download flows are in an disjoint region of the plot. This graph reveals that download and non-download flows have significantly different characteristics. A future traffic model has to distinguish between both types of traffic. In addition, the traffic profile provided in [191] gives evidence that the expected "mice and elephant" phenomenon in eDonkey traffic is not as severe as expected.

The feasibility of *mobile P2P* file sharing in infrastructure-based General Packet Radio Service (GPRS) and Universal Mobile Telecommunications System (UMTS) mobile networks is examined by measurements in [314]. The measurements have been carried out in networks of two German GPRS network operators and, for the first time, in an UMTS network. The subject of the empirical investigation was the eDonkey application, due to its hybrid architecture, which gives opportunities for network operators to interfere (cf. [315] and Section 6.4.2), and its continuing popularity [316]. The measurements demonstrate that mobile P2P is technically feasible for GPRS technology but stability and throughput are unacceptably low if compared to fixed P2P. Particularly, the direct exchange of large parts of files between two mobile peers and *multiple source download* (MSD) is not practical in GPRS. UMTS technology, in contrast, is more stable and has superior throughput, extending the capabilities of GPRS service into sufficient performance for mobile P2P file sharing. Figure 6.4(a) depicts the observed throughput for a MSD for a mobile peer downloading from a fixed peer and a mobile peer (abbreviated as fix/mob→mob). The throughput for MSD reaches a sustained of level of 25 KByte/s. This is a value which permits even the download of larger files.

The number of traversals of the air interface, however, has to be minimized in order to reduce the traffic and the transmission delay. Figure 6.4(b) compares the TCP connection holding times for an eDonkey file part of the same size for a fixed-to-mobile transmission and for a mobile-to-mobile transmission in UMTS. The figure reveals that the uplink capacity of the providing mobile peer is the bottleneck and that the connection setup time is almost doubled in the mobile-to-mobile transmission. A reduction of the necessary traversal of the air interface can be achieved efficiently by the application of caches, which has also the advantage of overcoming the asymmetric access bandwidths of mobile stations [317].

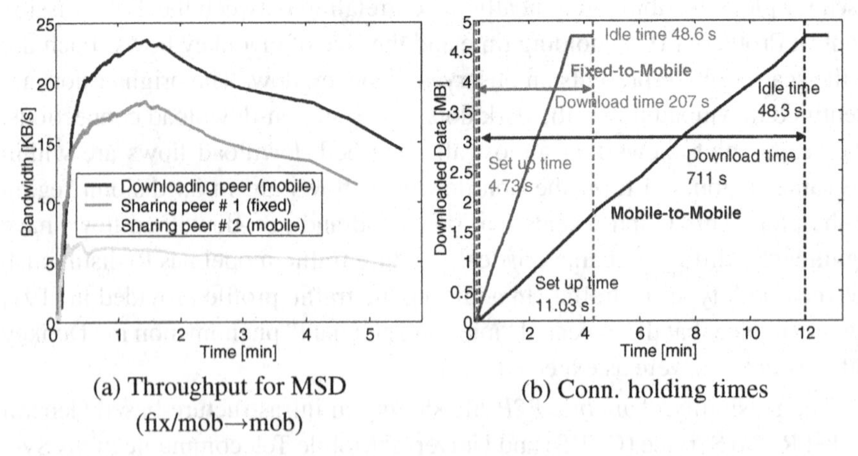

(a) Throughput for MSD (fix/mob→mob)

(b) Conn. holding times

Figure 6.4: Performance of mobile P2P file sharing in UMTS

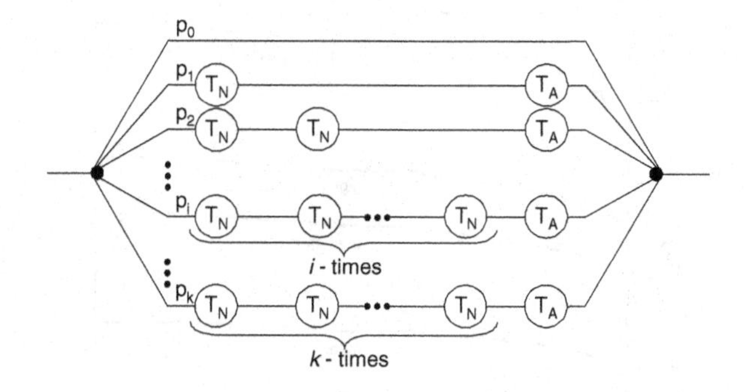

Figure 6.5: Phase diagram of the search duration T

6.4 Performance of P2P Mechanisms

6.4.1 P2P-Based Resource Mediation

The impact of network delay variation on resource mediation times, i.e., search times, in DHT based P2P systems is evaluated in [318]. The main goal is to prove scalability of Chord rings in order to guarantee Quality-of-Service (QoS) demands in large peer populations. Therefore, an analytical performance model for real-time applications based on the Chord algorithm is deduced. The phase diagram of the search delay is depicted in Figure 6.5.

A particular path i is chosen with probability p_i, where phase i consists of i network transmissions T_N to forward the query to the closest known finger and one network transmission T_A to send the answer back to the searching peer. By means of the phase diagram, the generating function and the Laplace transformation respectively can be derived to cope with the case of discrete-time or continuous-time network transfer delay. The distribution function of the search delay as seen from a user entering a search query to a peer in the Chord ring is computed. The analysis also gives insight into the quantiles of the search delay. Figure 6.6 shows different quantiles of the search delay. It can be seen that Chord searches indeed scale logarithmically. The curve with the 99%-quantile indicates that 99 percent of search durations lie below that curve. For a peer population of, e.g., $n = 3000$, in 99 percent of all cases the search delay is less then roughly 15 times the average network latency. That is, the curves indicate bounds of the search delay, which can be used for dimensioning purposes. Compared to the mean of the search delay, the quantiles of the search delay are on a significantly higher level. Still, the search delay scales in an analogous manner for the search delay quantiles.

The performance and stability of Chord rings under a Weibull lifetime dis-

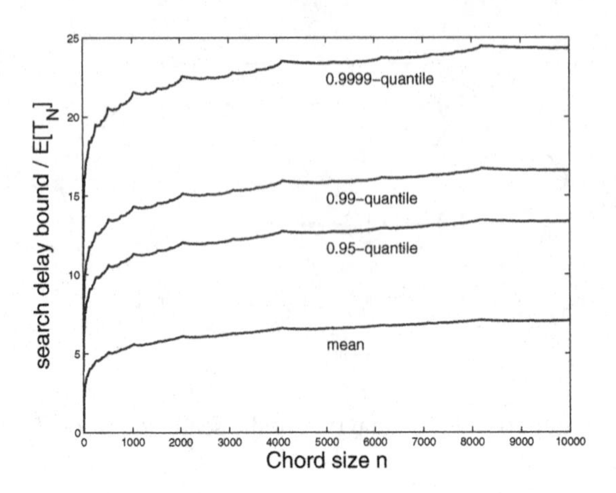

Figure 6.6: Search delay quantiles

tribution for modelling the *churn* behavior of peers is investigated in [319]. The churn behavior characterizes the ability of a node to join or leave arbitrarily, and is a key feature of P2P systems. The Weibull assumption qualitatively corresponds to measurements on some large real P2P systems [320] which show a heavy tail distribution of the life time. In previous studies only the exponential distribution was used. The latter suggest that a good level of performance of Chord systems can be achieved by dimensioning the frequency of maintenance messages per node with respect to the *half-life* or median of the distribution. In the case of skewed distributions, things can get more complicated and this approach may be suboptimal. It was shown in [319] that, in the Weibull case, the median of the distribution of the remaining life time can vary significantly depending on the update strategy. On the other hand, knowledge of each node's history in the system would lead to an individual half-life for each node. This could be used to obtain a closer to optimal maintenance strategy with less overhead. It seems that it is worth to study such strategies that allow gathering and sharing information on a node's character, or its *reputation.*

6.4.2 Influencing P2P User Data Traffic

A method to prioritize P2P transactions between peers connected to the same Internet Service Provider (ISP) network is presented in [321]. The method uses *Local Preference* and aims at the minimization of P2P traffic across domains boundaries, thus reducing the associated interconnection costs. In conjunction, it improves the QoS for the user by cutting the mean download time.

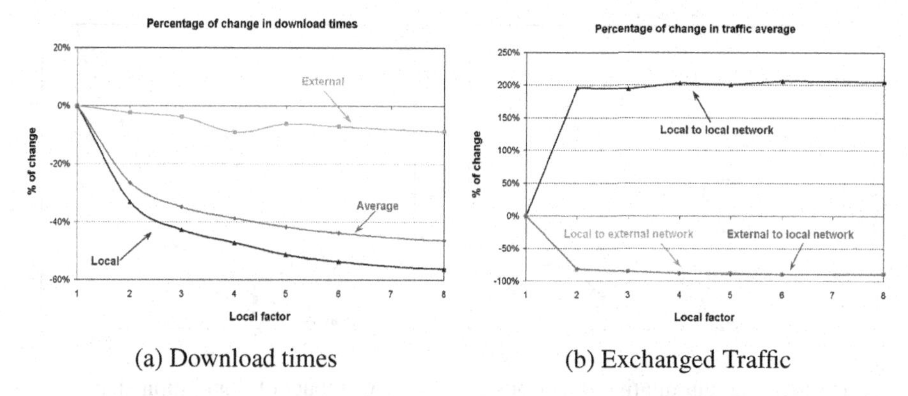

(a) Download times (b) Exchanged Traffic

Figure 6.7: Performance improvement by Local Preference in P2P file sharing systems

Local Preference is based on the community behavior and locality effects observed in P2P file sharing systems [322, 323, 324]. It assumes that most users in a given network are downloading the same files and that it is possible to encourage the users in a provider domain to exchange data mainly within their community. The investigated P2P file sharing system was eDonkey, however, the results are applicable to other P2P file sharing applications.

A main feature of eDonkey is the use of a distributed, priority queuing scheme for upload requests [325]. Each request is assigned a score value determining the priority for its service. The *Local Preference* is implemented as an additional "local factor" in the score equation. This enables the peer to decide locally whether to select this function or to disable it. This feature preserves the autonomy of the peers. The identification of local peers can be achieved in several different ways, e.g., by relating the peer's IP address to an ISP, Autonomous System (AS), or country using enhanced Domain Name Server entries [326].

The performance of *Local Preference* is depicted in Figure 6.7. Part (a) of the figure shows a significant improvement of up to 45% of the download times when using the locality mechanism. Figure 6.7(b) reveals that an increased consideration of the locality leads to significantly increased intra-network traffic ("local to local network") and reduces the transmitted data volume to other domains ("local to external network" or "external to local network") to a minimum.

6.4.3 A P2P Epidemic Algorithm for Content Distribution

Epidemic algorithms are efficient solutions for information dissemination in distributed settings. Their non-deterministic nature and peer-based communi-

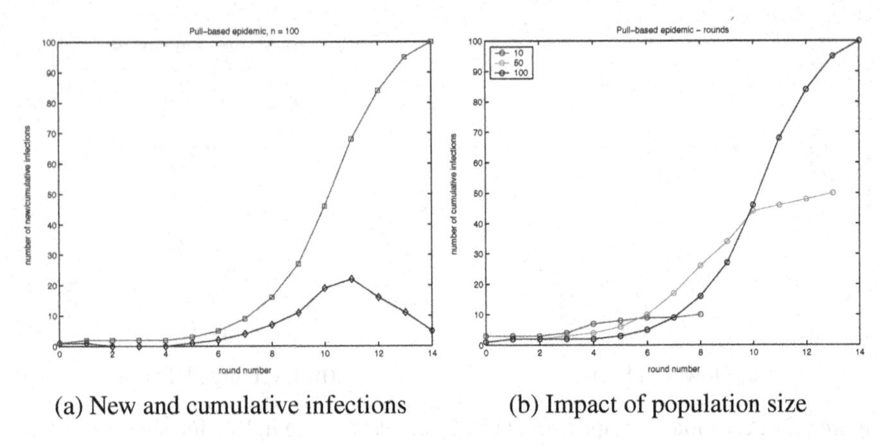

(a) New and cumulative infections (b) Impact of population size

Figure 6.8: Content Propagation

cation style are well matched to the requirements of ad hoc networks. There exists a number of recent studies that use epidemic mechanisms in ad hoc networks for the aim of improving multicast reliability [327, 328, 329]. The use of epidemic techniques, in particular the application of anti-entropy distribution model for reliable multicasting in ad hoc networks is investigated in [96]. In this study, a brief survey of the topic is presented, and formulations and analytical results for simple epidemics are described. Then, the P2P anti-entropy algorithm for content distribution is described together with initial results demonstrating the behavior of the algorithm.

Epidemic algorithms are based on the theory of epidemics and involve P2P propagation of updates. A popular distribution model based on the theory of epidemics is the anti-entropy, in which each peer chooses randomly another peer and exchanges its state information with it. In the case of state inconsistency between two peers, they exchange the updates using peer-to-peer connections. Further details of the algorithm, essential parameters, and simulation model are provided in [96].

As depicted in Figure 6.8(a), simulation results show that the distributions for new and cumulative content propagation are quite consistent when compared to expected values for simple epidemics. For different population sizes (n = 10, 50 and 100), the number of cumulative infections as a function of time is illustrated in Figure 6.8(b), showing scalability of content dissemination as the size of the population scales up. Variation of gossip rate parameter is also explored to observe its effect on the performance of content distribution. An increase in gossip rate results in an increase of the time needed for full infection. But, as a benefit, this decreases the average overhead per peer.

Furthermore, there exist issues such as membership, network awareness, buffer management, and message filtering that should be considered during the development of an epidemic information dissemination protocol [330]. In [331] the network awareness and buffer management issues are examined. In a large scale system, any epidemic algorithm that does not address actual network topology may lead to a dramatic increase in the underlying network load. Moreover, to ensure reliability, a loss recovery mechanism exploiting an efficient buffer management technique must be considered. Within this context, in [331] are described existing solutions to network awareness and buffer management issues as well as their approaches, and mathematical evaluation and comparative simulation results are given. Approaches presented in this study are easy to implement, fully distributed, and suitable to dynamic structures. The results show that they decrease network load dramatically compared with flat epidemic and the other clustered epidemic approaches.

List of Abbreviations

2G	Second Generation
3G	Third Generation
3GPP	Third Generation Partnership Project
A-DCH	Associated Dedicated Channel
ABA	Adaptive Bandwidth Allocation
ABC	Always Best Connected
ABR	Available Bit Rate
AC	Admission Control
ADSL	Asymmetric Digital Subscriber Line
AF	Assured Forwarding
AMC	Adaptive Modulation Coding
AMP	Adaptive Multi-Path Routing
AMS	Anick-Mitra-Sondhi
AP	Absolute Priority
AQM	Active Queue Management
AQUILA	Adaptive Resource Control for QoS Using an IP-based Layered Architecture
ARQ	Automatic Repeat Request
ASIC	Application-Specific Integrated Circuit
AS	Active Set
AS	Autonomous System
ATM	Asynchronous Transfer Mode

B-ISDN	Broadband Integrated Services Data Network
BB	Bandwidth Broker
B2B	Border-to-Border
BBB	Border-to-Border Budget
BGP	Border Gateway Protocol
BM	Bimodal Multicast
BMAP	Batch Markovian Arrival Process
BS	Base Station
BU	Burstification Unit
CAC	Connection Admission Control
CAI	Continue after Interruption
CAIDA	Cooperative Association for Internet Data Analysis
CBR	Constant Bit Rate
CCD	Complementary Cumulative Distribution
CDMA	Code Division Multiple Access
CLT	Central Limit Theorem
CO	Capacity Overprovisioning
CRC	Cyclic Redundancy Check
CSMA	Carrier Sense Multiple Access
CTMC	Continuous-Time Markov Chain
CoV	Coefficient of Variation
DB	Delay Budget
DCF	Distributed Coordination Function
DCH	Dedicated Channel
DDR	Decompose-Design-Reassemble
DHR	Decreasing Hazard Rate
DHT	Distributed Hash Table
DiffServ	Differentiated Services
DLC	Data Link Control

DLL	Data Link Layer
DPMI	Distributed Passive Measurement Infrastructure
DRR	Deficit Round-Robin
DSCH	Downlink Shared Channel
EB	Egress Budget
ECMP	Equal Cost Multi-Path
ECN	Explicit Congestion Notification
EF	Expedited Forwarding
ELB	Egress Link Budget
EM	Expectation-Maximization
EMW	Elwalid-Mitra-Wentworth
EuroNGI	Design and Engineering of the Next Generation Internet
F-RTO	Forward-Retransmission Timeout Recovery
FACH	Forward Access Channel
FB	Foreground-Background
FBM	Fractional Brownian Motion
FCFS	First-Come-First-Served
FDL	Fiber Delay Line
FEC	Forward Error Correction
FFM	Fluid Flow Model
FH	Frequency Hopping
FHSS	Frequency Hopping Spread Spectrum
FIFO	First-In-First-Out
FPI	First Step of the Policy Iteration
FTP	File Transfer Protocol
FTDMA	Fixed Time-Division Multiple Access
FXC	Fiber Switching Cross-Connect
GFA	Gateway Foreign Agent
GMPLS	Generalized Multi-Protocol Label Switching

GPRS General Packet Radio Service

GPS Generalized Processor Sharing

GPS Global Positioning System

GSM Global System for Mobile

HA Home Agent

HMM Hidden Markov Model

HOL Head-of-Line

HOL-PJ Head-of-Line with Priority Jumps

HS-DSCH High-Speed Downlink Shared Channel

HSDPA High-Speed Downlink Packet Access

HTTP HyperText Transfer Protocol

IB Ingress Budget

IB/EB Ingress and Egress Budget

IDAC Inter-Domain Admission Control

IETF Internet Engineering Task Force

iid Independent and Identically Distributed

ILB Ingress Link Budget

ILB/ELB Ingress and Egress Link Budget

ILP Integer Linear Programming

IP Internet Protocol

IPDV IP Packet Delay Variation

IS-IS Intermediate System to Intermediate System

ISL Inter Satellite Link

ISP Internet Service Provider

IST Information Society Technologies

JET Just Enough Time

JIT Just-in-Time

KING Key components for the Internet of the Next Generation

LAC Link Admission Control

LAN	Local Area Network
LAS	Least-Attained Service
LAUC	Latest Available Unscheduled Channel
LAUC-VF	Latest Available Unscheduled Channel with Void-Filling
LA	Location Area
LB	Link Budget
LCA	Linear to Consumer Activity
LEO	Low Earth Orbit
LISDLG	Limit of the Integrated Superposition of Diffusion Processes with Linear Differential Generator
LLC	Logic Link Control
LPC	Linear Predictive Coding
LPCA	Linear to Provider and Consumer Activity
LP	Linear Programming
LRD	Long Range Dependence/Dependent
LSP	Label Switched Path
MAC	Medium Access Control
MAI	Multiple Access Interference
MAr	Measurement Area
MArN	Measurement Area Network
MBAC	Measurement-Based Admission Control
MDP	Markov Decision Process
MEDF	Modified Earliest Deadline First
MEMS	Micro-Electro-Mechanical System
MG-OXC	Multi-Granular Optical Cross-Connect
MHD	Multi-Hour Design
MHS	Message Handling Service
MIP	Mobile IP
MLPS	Multilevel Processor Sharing
MMFP	Markov Modulated Fluid Process

MMPP Markov Modulated Poisson Process
MMRP Markov Modulated Rate Process
MMS Multimedia Messaging Service
MN Mobile Node
MPEG Moving Pictures Expert Group
MPLS Multi-Protocol Label Switching
MP Measurement Point
MRDV Multipath Routing with Dynamic Variance
MS Mobile Station
MSD Multiple Source Download
MTU Maximum Transmission Unit
NAC Network Admission Control
NAT Network Address Translator
nFA new Foreign Agent
NGN Next Generation Network
NJ Negligible Jitter
NRT Non Real-Time
NTP Network Time Protocol
OBS Optical Burst Switching
OD Origin-Destination
ODE Ordinary Differential Equation
O/E/O Optical/Electronic/Optical
oFA old Foreign Agent
OPS Optical Packet Switching
OSNR Optical Signal-to-Noise Ratio
OSPF Open Shortest Path First
OTDM Optical Time Division Multiplexing
OUP Ornstein-Uhlenbeck Process
OXC Optical Cross-Connect

p-RAI	Partial Repeat after Interruption
P2P	Peer-to-Peer
PCAP	Packet CAPture
PCBR	Premium Constant Bit Rate
pdf	Probability Density Function
PDU	Protocol Data Unit
PF	Perron-Frobenius
PFQ	Priority Fair Queueing
PFS	Priority Forcing Scheme
PGF	Probability Generating Function
PH-type	Phase-Type
PLR	Packet Loss Ratio
PLRG	Power-Law Random Graph
PMC	Premium Mission Critical
PMF	Probability Mass Function
PMM	Premium Multi-Media
POP	Point of Presence
PP	Path Protection
PQ	Priority Queueing
PSTN	Public Switched Telephone Network
PS	Processor Sharing
PVBR	Premium Variable Bit Rate
PVC	Permanent Virtual Circuit
QBD	Quasi-Birth-Death
QoS	Quality of Service
RAI	Repeat after Interruption
RAM	Random Access Memory
RB	Standby-Random Backoff
RC	Resource Controller

RED	Random Early Detection
REM	Rate Envelope Multiplexing
RER	Random Early Reroute
RFC	Request for Comments
RKHS	Reproducing Kernel Hilbert Space
RMON	Remote Monitoring
RNC	Radio Network Controller
RSM	Rate Sharing Multiplexing
RSR	Reduced Service Rate
RSVP	ReSerVation Protocol
RT	Real-Time
RT/NRT	Real-Time/Non-Real-Time
RTP	Real Time Protocol
RTS-CTS	Ready-to-Send-Clear-to-Send
RTT	Round-Trip Time
RWA	Routing and Wavelength Assignment
RWP	Random Waypoint Model
S-STPT	Self-Synchronized Time Packet Transmission
SACK	Selective Acknowledgement
SB	Standby-Random Backoff
SBA	Static Bandwidth Allocation
semi-BD	Semi-Birth-and-Death
SFQ	Start-time Fair Queueing
SIMA	Simple Integrated Media Access
SIR	Signal-to-Interference-Ratio
SLA	Service Level Agreement
SMA	Super Measurement Area
SNMP	Simple Network Management Protocol
SNR	Signal-to-Noise Ratio

SPM Self-Protecting Multi-path

SRM Scalable Reliable Multicast

SRR Surplus Round-Robin

SiRM Simplified Reference Model

TCD Traffic Class Dependent

TCP Transmission Control Protocol

TD Temporary Document

TDAC Traffic Descriptor-based Admission Control

TM Traffic Matrix

tMAP Transient Markovian Arrival Process

UDP User Datagram Protocol

UMTS Universal Mobile Telecommunications System

VBR Variable Bit Rate

VoIP Voice over IP

WAN Wide Area Network

WAP Wireless Application Protocol

WBXC Wave-Band Switching Cross-Connect

WCDMA Wideband Code Division Multiple Access

WFQ Weighted Fair Queueing

WIXC Wavelength Interchanging Cross-Connect

WLAN Wireless Local Area Network

WRR Weighted Round-Robin

WSXC Wavelength Selective Cross-Connect

WVPing Wave-Ping

WWW World Wide Web

xDSL x Digital Subscriber Line

Bibliography

[1] J. Charzinski, C. Hoogendoorn, K. Schrodi, C. Winkler, and M. N. Huber. Towards Carrier-Grade Next Generation Networks. Technical Report 279TD(03)040, COST-279, 2003. [cf. COST-279, TD(03)040]. 6, 24

[2] A. Bak, W. Burakowski, F. Ricciato, S. Salsano, and H. Tarasiuk. Traffic handling in AQUILA QoS IP networks, Quality of Future Internet Services. In M. I. Smirnov, J. Crowcroft, J. Roberts, and F. Boavida, editors, *Quality of Future Internet Services: Second COST 263 International Workshop, Qofis 2001, Coimbra, Portugal, September 24-26, 2001, Proceedings*, volume 2156/2001 of *Lecture Notes in Computer Science*, pages 243–260. Springer-Verlag Heidelberg, January 2001. [cf. COST-279, TD(01)017]. 8, 13, 20

[3] K. Lindberger. Dimensioning and design methods for integrated ATM networks. In *Proc. of 14-th International Teletraffic Congress, Antibes, 1994*, 1994. 9

[4] C. Brandauer and P. Dorfinger. An implementation of a service class providing assured TCP rates with the AQUILA framework. In W. Burakowski, B. F. Koch, and A. Beben, editors, *Architectures for Quality of Service in the Internet 2003*, number 2698 in Lecture Notes in Computer Science, 2003. 9

[5] W. Burakowski and H. Tarasiuk. Admission control for TCP connections in QoS IP network. In C. Chung, C. Kim, W. Kim, T. Ling, and K. Song, editors, *Web and Communication Technologies and Internet-Related Social Issue—HSI 2003: Second International Conference on Human.Society@INTERNET, Seoul, Korea, June 2003, Proceedings*, volume 2713/2003 of *Lecture Notes in Computer Science*, pages 283–293. Springer-Verlag Heidelberg, January 2003. [cf. COST-279, TD(02)029]. 9, 16, 17

[6] M. Dabrowski, G. Eichler, M. Fudala, D. Katzengruber, T. Kilkanen, N. Miettinen, H. Tarasiuk, and M. Titze. Evaluation of the AQUILA architecture: Trial results for signalling performance, network services and user acceptance. In W. Burakowski, B. F. Koch, and A. Beben, editors, *Architectures for Quality of Service in the Internet 2003*, number 2698 in Lecture Notes in Computer Science, 2003. 9

[7] E. Nyberg and S. Aalto. How to achieve fair differentiation. In *Proceedings of Networking 2002*, pages 1178–1183, Pisa, Italy, May 2002. IFIP-TC6, Springer-Verlag. [cf. COST-279, TD(01)004], [332, 333]. 9, 222

[8] K. Kilkki and J. Ruutu. Simple Integrated Media Access—an Internet service based on priorities. In *6th International Conference on Telecommunication Systems*, 1998. 9

[9] H. Tarasiuk, W. Burakowski, and R. Janowski. On Assuring End-to-End QoS in Heterogeneous Networks by Investigating Network Service Concept. Technical Report 279TD(05)014, COST-279, 2005. [cf. COST-279, TD(05)014]. 10, 11

[10] K. H. Chan and J. Babiarz. Inter-provider service classes. In *Inter-provider QoS Meeting*, January 2005. 10

[11] H. Bruneel and B. G. Kim. *Discrete-Time Models for Communication Systems Including ATM.* Kluwer Academic Publishers, 1992. 11

[12] A. Kortebi, S. Oueslati, and J. Roberts. Cross-protect: Implicit service differentiation and admission control. In *IEEE HPSR*, 2004. [cf. COST-279, TD(04)003]. 11, 41

[13] P. Goyal, H. Vin, and H. Cheng. Start-time Fair Queueing: A scheduling algorithm for Integrated Services packet switching networks. *IEEE/ACM Trans. on Networking*, 5(5):690–704, October 1997. 11

[14] A. Kortebi, L. Muscariello, S. Oueslati, and J. Roberts. Evaluating the number of active flows in a scheduler realizing fair statistical bandwidth sharing. In *ACM Sigmetrics 2005*, Banff, Alberta, Canada, June 2005. ACM. [cf. COST-279, TD(05)006]. 12, 41

[15] R. Janowski and K. Rzepakowski. Estimation of the admissible load in a two priority system. In Peter Buchholz, Ralf Lehnert, and Michal

Pióro, editors, *MMB*, pages 225–234. VDE Verlag, September 2004. [cf. COST-279, TD(04)030]. 13, 79

[16] H. Tuan Tran, T. Ziegler, and F. Ricciato. QoS provisioning for VoIP traffic by deploying admission control. In *LNCS 2698, Proceedings of Workshop on Architectures for Quality of Service in the Internet, Art-QoS*, pages 139–153, March 2003. [cf. COST-279, TD(02)037], [334]. 14, 77, 222

[17] T. Bonald, A. Proutière, and J. W. Roberts. Statistical performance guarantees for streaming flows using Expedited Forwarding. In *Proceedings of IEEE INFOCOM*, volume 2, 2001. 14

[18] M. Dabrowski and W. Burakowski. Assessment of Token Bucket parameters by on-line traffic measurements. In *10th Polish Teletraffic Symposium*, September 2003. [cf. COST-279, TD(03)044]. 15, 113

[19] A. Elwalid, D. Mitra, and R. H. Wentworth. A new approach for calculating buffers and bandwidth to heterogeneous, regulated traffic in an ATM node. *IEEE Journal on Selected Areas in Communications*, 13(6), August 1995. 15

[20] M. Menth and O. Rose. Performance tradeoffs for header compression in MPLS networks. In *10th International Telecommunication Network Planning Symposium*, pages 503–508, Munich, Germany, 2002. [cf. COST-279, TD(02)031]. 15, 16

[21] W. Burakowski and A. Beben. Premium Message Handling service in IP QoS networks. In *10th Polish Teletraffic Symposium*. Publishing House of Electrical Engineering Faculty, September 2003. [cf. COST-279, TD(03)035]. 17

[22] A. Beben and R. Janowski. Statistical admission control for WWW Traffic. In *10th Polish Teletraffic Symposium*. Publishing House of Electrical Engineering Faculty, September 2003. [cf. COST-279, TD(03)045]. 17

[23] J. Padhye, V. Firoiu, D. F. Towsley, and J. F. Kurose. Modeling TCP Reno performance: a simple model and its empirical validation. *IEEE/ACM Trans. on Networking*, 8(2):133–145, 2000. 18, 36

[24] N. Benameur, S. Oueslati, and J. W. Roberts. Experimental Implementation of Implicit Admission Control. Technical Report 279TD(03)026, COST-279, 2003. [cf. COST-279, TD(03)026]. 18

[25] S. Ben Fredj, S. Oueslati-Boulahia, and J.W. Roberts. Measurement-based Admission Control for Elastic Traffic. In J. Moreira de Souza, N. L. S. da Fonseca, and E. A. de Souza e Silva, editors, *Teletraffic Engineering in the Internet Era*, pages 161–172. ITC 17, Elsevier, December 2001. 18

[26] A. Terzis, J. Wang, J. Ogawa, and L. Zhang. A two-tier resource management model for the Internet. In *Global Internet Symposium'99*, December 1999. 19

[27] M. Menth, S. Kopf, J. Milbrandt, and J. Charzinski. Introduction to budget based network admission control methods. In 28th *Annual IEEE Conference on Local Computer Networs (LCN2003)*, Bonn, Germany, October 2003. 19

[28] M. Menth. *Efficient Admission Control and Routing in Resilient Communication Networks*. PhD thesis, University of Würzburg, Faculty of Computer Science, Am Hubland, July 2004. http://opus.bibliothek.uni-wuerzburg.de/opus/volltexte/2004/994/. 19, 24

[29] M. Menth, S. Gehrsitz, and J. Milbrandt. Fair assignment of efficient network admission control budgets. In 18th *International Teletraffic Congress*, pages 1121–1130, Berlin, Germany, September 2003. 19

[30] X. Xiao and L. M. Ni. Internet QoS: A big picture. *IEEE Network Magazine*, 13(2):8–18, March 1999. 20

[31] N. G. Duffield, P. Goyal, A. G. Greenberg, P. P. Mishra, K. K. Ramakrishnan, and J. E. van der Merive. A flexible model for resource management in Virtual Private Networks. In *SIGCOMM*, pages 95–108, 1999. 20

[32] M. Menth, S. Kopf, and J. Milbrandt. A performance evaluation framework for network admission control methods. In *IEEE Network Operations and Management Symposium (NOMS)*, Seoul, South Korea, April 2004. 20

[33] C. Mauz. Mapping of arbitrary traffic demand and network topology on a mesh of rings network. In *Proceedings of the Fifth Working Conference on Optical Network Design and Modelling (ONDM'01)*, February 2001. 20

[34] M. Menth and J. Charzinski. Impact of network topology on the performance of network admission control methods. In *International Work-*

shop on Multimedia Interactive Protocols and Systems (MIPS2003), pages 195–206, Napoli, Italy, November 2003. 20

[35] M. Menth, J. Milbrandt, and S. Kopf. Impact of routing and traffic distribution on the performance of network admission control. In 9th*IEEE Symposium on Computers and Communications (ISCC)*, pages 883–890, Alexandria, Egypt, June 2004. 20

[36] G. A Politis, P. Sampatakos, and I. S. Venieris. Design of a multilayer Bandwidth Broker architecture. In *Interworking*, Bergen, Norway, 2000. 21

[37] T. Engel, E. Nikolouzou, F. Ricciato, and P. Sampatakos. Analysis of adaptive resource distribution algorithms in the framework of a dynamic DiffServ IP network. In 8th *International Conference on Advances in Communications and Control (ComCon8)*, Crete, Greece, June 2001. 21

[38] M. Menth. A scalable protocol architecture for end-to-end signaling and resource reservation in IP networks. In *17th International Teletraffic Congress*, pages 211–222, Salvador da Bahia, Brazil, November 2001. [cf. COST-279, TD(01)010]. 21

[39] M. Menth, J. Milbrandt, and S. Kopf. Adaptive Bandwidth Allocation for Wide Area Networks. Technical Report 279TD(05)021, COST-279, 2005. [cf. COST-279, TD(05)021]. 23

[40] M. Menth, S. Kopf, and J. Charzinski. Network admission control for fault-tolerant QoS provisioning. In *HSNMC*, pages 1–13, Toulouse, France, June 2004. [cf. COST-279, TD(03)029]. 24, 25

[41] M. Menth, A. Reifert, and J. Milbrandt. Self-protecting multipaths—a simple and resource-efficient protection switching mechanism for MPLS networks. In 3rd*IFIP-TC6 Networking Conference (Networking)*, pages 526–537, Athens, Greece, May 2004. [cf. COST-279, TD(03)046], [335]. 25, 49, 223

[42] R. Martin, M. Menth, and J. Charzinski. Comparison of border-to-border budget based network admission control and capacity overprovisioning. In 4th*IFIP-TC6 Networking Conference (Networking)*, Waterloo, Canada, May 2005. [cf. COST-279, TD(05)020]. 26

[43] R. Martin, M. Menth, and J. Charzinski. Comparison of Link-by-Link Admission Control and Capacity Overprovisioning. In *submitted to* 19th*International Teletraffic Congress*, 2005. 26

[44] M. Dabrowski, A. Beben, and W. Burakowski. On inter-domain admission control supported by measurements in multi-domain IP QoS network. In *IPS 2004 Inter-Domain Performance and Simulation*, March 2004. [cf. COST-279, TD(04)016]. 28, 113

[45] ITU-T. Recommendation Y.1540 (2002), Internet Protocol Data Communication Service—IP Packet transfer and Availability Performance Parameters, December 2002. 28

[46] E. J. Chen and W. D. Kelton. Quantile and histogram estimation. In *WSC*, pages 95–108, Arlington, USA, 2001. 28

[47] P. Olivier and N. Benameur. Flow level IP traffic characterization. In *17th International Teletraffic Congress*, Salvador da Bahia, Brazil, December 2001. [cf. COST-279, TD(01)014]. 29, 100, 103

[48] J.W. Cohen. The multiple phase service network with Generalized Processor Sharing. *Acta Informatica*, 12:245–284, 1979. 30, 34, 72, 73, 104, 105

[49] L. Kleinrock. *Queueing Systems, Volume 2: Computer Applications*. John Wiley & Sons, New York, 1976. 30, 75

[50] T. Bonald and A. Proutière. Insensitive bandwidth sharing in data networks. *Queueing Systems: Theory and Applications*, 44(1):69–100, 2003. [cf. COST-279, TD(04)014], [53]. 30, 194

[51] L. Massoulié and J.W. Roberts. Bandwidth sharing: Objectives and algorithms. *IEEE/ACM Trans. on Networking*, 10(3):320–328, 2002. 30

[52] R. F. Serfozo. *Introduction to Stochastic Networks*. Springer, 1999. 30

[53] T. Bonald and A. Proutière. On performance bounds for Balanced Fairness. *Performance Evaluation*, 55(1-2):25–50, 2004. [cf. COST-279, TD(04)014], [50]. 30, 194

[54] T. Bonald, P. Olivier, and J. Roberts. Dimensioning high speed IP access networks. In *18th International Teletraffic Congress*, Berlin, Germany, September 2003. [cf. COST-279, TD(03)007], [336]. 31, 223

[55] D. P. Heyman, T. V. Lakshman, and A. L. Neidhardt. A new method for analysing feedback-based protocols with applications to engineering Web traffic over the Internet. In *Proceedings of ACM Sigmetrics '97*, 1997. 31

[56] V. Baek Iversen and T. Holmberg. Resource Sharing Models for Quality-of-Service. Technical Report 279TD(04)037, COST-279, 2004. [cf. COST-279, TD(04)037]. 31, 75

[57] H. van den Berg, M. Mandjes, R. van de Meent, A. Pras, F. Roijers, and P. Venemans. QoS Aware Bandwidth Provisioning in IP Backbone Networks. Technical Report 279TD(03)034, COST-279, 2003. [cf. COST-279, TD(03)034]. 32, 33, 87, 106

[58] S. Ben Fredj, T. Bonald, A. Proutière, G. Régnié, and J.W. Roberts. Statistical bandwidth sharing: a study of congestion at flow level. In *Proceedings of ACM SIGCOMM*, San Diego, CA, USA, August 2001. 32, 74

[59] R. Vranken, R. van der Mei, R. E. Kooij, and H. van den Berg. Flow-level performance models for the TCP with QoS differentiation. In *Proceedings of the International Seminar on Telecommunication Networks and Teletraffic Theory*, pages 78–87, St. Petersburg, Russia, 2002. [cf. COST-279, TD(02)010]. 33, 74, 103, 104

[60] R.D. van der Mei, J.L. van den Berg, R. Vranken, and B.M.M. Gijsen. Sojourn times in multi-server Processor Sharing systems with priorities. *Performance Evaluation*, 54:249–261, 2003. 33

[61] R. D. van der Mei, H. van den Berg, R. Vranken, and B. Gijsen. Analysis of a flow level model for TCP behavior in case of Priority Queueing. Technical Report 279TD(01)012, COST-279, 2001. [cf. COST-279, TD(01)012]. 34, 73, 103, 104

[62] F. Delcoigne, A. Proutière, and G. Régnié. Modelling integration of streaming and data traffic. In *15th ITC Specialist Seminar on Internet Traffic Engineering and Traffic Management*, Würzburg, Germany, July 2002. [cf. COST-279, TD(02)019], [337]. 35, 74, 104, 223

[63] A. de Vendictis and A. Baiocchi. Investigating TCP single source behavior in time-varying capacity network scenarios. In *18th International Teletraffic Congress*, Berlin, Germany, August/September 2003. [cf. COST-279, TD(03)012]. 35

[64] T. V. Lakshman and U. Madhow. The performance of TCP/IP for networks with high bandwidth-delay products and random loss. *IEEE/ACM Trans. on Networking*, 5(3):336–350, 1997. 36

[65] A. Jena and A. Popescu. Traffic engineering for Internet applications. In *Internet Performance and Control of Network Systems II*, volume 4523, pages 67–78, Denver, USA, August 2001. SPIE. [cf. COST-279, TD(01)007]. 37

[66] S. Floyd. HighSpeed TCP for large congestion window. Technical Report RFC 3649, IETF, December 2003. 38

[67] C. Jin, D. X. Wei, and S. H. Low. FAST TCP: motivation, architecture, algorithms, performance. In *Proceedings of IEEE Infocom*, March 2004. 38, 39

[68] D. Katabi, M. Handley, and C. Rohrs. Internet congestion control for high bandwidth-delay product networks. In *Proceedings of ACM Sigcomm 2002*, August 2002. 38

[69] Tom Kelly. Scalable TCP: Improving performance in high-speed wide area networks. *ACM SIGCOMM Computer Communication Review*, 33(2):83–91, April 2003. 38

[70] E. Souza and D. Agarwal. A HighSpeed TCP study: Characteristics and deployment issues. Technical report, Lawrence Berkeley National Lab, Berkeley, USA, 2003. 38

[71] S. Molnár, T. Anh Trinh, and B. Sonkoly. A Performance Study of High Speed TCP. Technical Report 279TD(04)012, COST-279, 2004. [cf. COST-279, TD(04)012]. 38

[72] M. J. Osborne and A. Rubenstein. *A Course in Game Theory*. The MIT Press, Cambridge, Massachusetts, 1994. 39

[73] S. Molnár and T. Anh Trinh. Analysis of TCP Vegas and FAST TCP in a Game-Theoretical Framework. Technical Report 279TD(05)016, COST-279, 2005. [cf. COST-279, TD(05)016]. 39

[74] D. Z. Lenardic, B. Zovko-Cihlar, and M. Grgic. Analysis of Network Buffering Effects on TCP/IP Protocol Behavior. Technical Report 279TD(03)014, COST-279, 2003. [cf. COST-279, TD(03)014]. 40

[75] S. Floyd and V. Jacobson. Random Early Detection gateways for congestion avoidance. *IEEE/ACM Trans. on Networking*, 1(4):397–413, 1993. 40

[76] E. Plasser, T. Ziegler, and P. Reichl. On the non linearity of the RED drop function. In *ICCC*, Mubai, India, August 2002. [cf. COST-279, TD(02)032]. 40

[77] M. Klimo and K. Bachratá. Impact of a Packet Loss Process to a Speech Process. Technical Report 279TD(05)001, COST-279, 2005. [cf. COST-279, TD(05)001], [338, 339, 340]. 40

[78] M. Nord, J. Soler Lucas, and V. Baek Iversen. Simulation and performance analysis of a GMPLS Lambda scheduler. In *The Internet Protocol and Optical Networking Workshop*, Grasmere, UK, September 2002. [cf. COST-279, TD(03)006]. 41

[79] R. Boël and S. de Vuyst. Prediction Based Resource Allocation, a Simulation Experiment. Technical Report 279TD(02)008, COST-279, 2002. [cf. COST-279, TD(02)008]. 42, 112

[80] R. Martin, M. Menth, and V. Phan. Performance of TCP/IP with MEDF scheduling. In 3rd*International Workshop on Quality of Future Internet Services (QofIS)*, pages 94–103, Barcelona, Spain, September 2004. [cf. COST-279, TD(04)011]. 42

[81] W. Burakowski and M. Fudala. Priority Forcing Scheme: a new strategy for getting better than Best Effort service in IP-based network. In *Internet Technologies, Applications and Societal Impact, Kluwer Academic Publishers*, Wroclaw, Poland, October 2002. [cf. COST-279, TD(02)039]. 42

[82] M. Ajmone Marsan, M. Franceschinis, E. Leonardi, F. Neri, and A. Tarello. Instability phenomena in underloaded packet networks with QoS schedulers. In *INFOCOM*, San Francisco, USA, March/April 2003. [cf. COST-279, TD(02)027], [341]. 43, 223

[83] A. Švigelj, M. Mohorcic, and G. Kandus. Analysis of Different Scheduling Policies in Traffic Class Dependent Routing for LEO Satellite Networks with ISLs. Technical Report 279TD(04)004, COST-279, 2004. [cf. COST-279, TD(04)004]. 43, 143

[84] F.-J. Ramón, J. Enríquez, J. Andrés, and A. Molínes. Multipath Routing with Dynamic Variance. Technical Report 279TD(02)043, COST-279, 2002. [cf. COST-279, TD(02)043]. 45, 47

[85] I. Gojmerac, T. Ziegler, F. Ricciato, and P. Reichl. Adaptive multipath routing for dynamic traffic engineering. In *IEEE Globecom 2003*, San Francisco, USA, 2003. [cf. COST-279, TD(03)028]. 45, 47

[86] J. Milbrandt, D. Staehle, S. Köhler, and L. Berry. Decomposition of Large IP Networks for Routing Optimization. Technical Report 279TD(02)005, COST-279, 2002. [cf. COST-279, TD(02)005]. 46

[87] S. Köhler and A. Binzenhöfer. MPLS traffic engineering in OSPF networks—a combined approach. In *ITC 18*, Berlin, Germany, August/September 2003. [cf. COST-279, TD(03)019]. 47

[88] R. Susitaival and S. Aalto. Providing Differentiated Services by Load Balancing and Scheduling in MPLS Networks. Technical Report 279TD(03)003, COST-279, 2003. [cf. COST-279, TD(03)003]. 47

[89] R. Susitaival, J. Virtamo, and S. Aalto. Load balancing by MPLS in Differentiated Services networks. In *International Workshop Art-Qos*, pages 252–264, Warsaw, Poland, March 2003. [cf. COST-279, TD(02)041]. 47

[90] N. Akar, I. Hökelek, M. Atik, and E. Karasan. A reordering-free multipath traffic engineering architecture for DiffServ/MPLS networks. In *IEEE Workshop on IP Operations and Management*, Kansas City, USA, October 2003. [cf. COST-279, TD(02)026]. 47

[91] O. Alparslan, N. Akar, and E. Karasan. Combined use of prioritized AIMD and flow-based traffic splitting for robust TCP load balancing. In *Proceedings of Workshop on Quality of Future Internet Services (QoFIS), LNCS Vol. 3266*, 2004. [cf. COST-279, TD(04)020], [342]. 48, 223

[92] J. Karvo and S. Aalto. Using Multicast or a Combination of Unicast and Broadcast for Transmitting Popular Content. Technical Report 279TD(03)020, COST-279, 2003. [cf. COST-279, TD(03)020]. 50

[93] Ö. Özkasap and M. Çaglar. Traffic behavior of Scalable Multicast: self-similarity and protocol dependence. In *18th International Teletraffic Congress*, Berlin, Germany, August/September 2003. [cf. COST-279, TD(02)038]. 51, 93

[94] Ö.Özkasap and M. Çaglar. Traffic characterization of Scalable Multicasting in the case of a self-similar source (Poster). In *ACM SIGCOMM*, Karlsruhe, Germany, August 2003. [cf. COST-279, TD(03)036], [343]. 51, 93, 94, 223

[95] Ö. Özkasap. Scalability and robustness of pull-based anti-entropy distribution model. In *ISCIS XVIII (18th International Symposium on Computer and Information Sciences)*, Antalya, Turkey, November 2003. [cf. COST-279, TD(03)050]. 51

[96] Z. Genç and Ö. Özkasap. Peer-to-Peer epidemic algorithms for reliable multicasting in ad-hoc networks. In *ICIT 2004: International Conference on Information Technology*, Istanbul, Turkey, December 2004. [cf. COST-279, TD(04)039]. 51, 176

[97] H. Reittu and I. Norros. On the effect of very large nodes in Internet graphs. In *Globecom 2002*, Taipei, Taiwan, 2002. [cf. COST-279, TD(02)002]. 52

[98] H. Reittu and I. Norros. On the power law random graph model of massive data networks. *Performance Evaluation*, 55(1-2):3–23, 2004. [cf. COST-279, TD(02)020]. 52, 53

[99] R. van der Hofstad, G. Hooghiemstra, and P. Van Mieghem. Random graphs with finite variance degrees. *Random Structures and Algorithms*, 2005. To appear. 53

[100] R. van der Hofstad, G. Hooghiemstra, and D. Znamenski. Distances in random graphs with finite mean and infinite variance degrees. 2005. Preprint. 53

[101] R. van der Hofstad, G. Hooghiemstra, and D. Znamenski. Distances in random graphs with infinite mean degrees. 2005. Submitted for publication to Extremes. 53

[102] I. Norros and H. Reittu. On a Conditionally Poissonian Graph Process. Technical Report 279TD(04)005, COST-279, 2004. [cf. COST-279, TD(04)005]. 53

[103] F. Chung and L. Lu. The average distance in a random graph with given expected degrees. *Internet Mathematics*, 1:91–114, 2003. 53

[104] I. Norros and H. Reittu. Architectural features of the power-law random graph model of Internet: notes on soft hierarchy, vulnerability

and multicasting. In *ITC-18*, Berlin, September 2003. [cf. COST-279, TD(03)016]. 53, 54

[105] J. Walraevens, B. Steyaert, and H. Bruneel. A single-server queue with a Priority Scheduling discipline: Performance study. Technical Report 279TD(01)006, COST-279, 2001. [cf. COST-279, TD(01)006]. 56, 57

[106] J. Walraevens, B. Steyaert, and H. Bruneel. Performance analysis of a GI-G-1 preemptive resume priority buffer. In *Proceedings of the Networking 2002 Conference, Pisa, Italy, May 19-24*, LNCS 2345, pages 745–756, 2002. 57

[107] J. Walraevens, B. Steyaert, and H. Bruneel. Analysis of a preemptive repeat priority buffer with resampling. In *Proceedings of the International Network Optimization Conference (INOC 2003), Evry, France, October 27-29*, pages 581–586, 2003. 57

[108] J. Walraevens, B. Steyaert, and H. Bruneel. Delay characteristics in discrete-time GI-G-1 queues with non-preemptive priority queueing discipline. *Performance Evaluation*, 50(1):53–75, 2002. 57

[109] D. Fiems, B. Steyaert, and H. Bruneel. Discrete-time queues with general service times and general server interruptions. In *SPIE*, volume 4211, pages 93–104, Boston, USA, 2000. [cf. COST-279, TD(01)005], [344]. 57, 58, 223

[110] T. Maertens, J. Walraevens, and H. Bruneel. Performance analysis of a single-server queue with HOL-PJ Priority Scheduling discipline. In *Proceedings of the Second International Working Conference on Performance Modelling and Evaluation of Heterogeneous Networks (HET-NETs '04)*, pages P42/1–P42/10, Ilkley, UK, July 2004. [cf. COST-279, TD(04)018]. 58, 59

[111] T. Maertens, J. Walraevens, and H. Bruneel. A discrete-time queue with a modified HOL priority scheduling discipline: performance analysis. In *Proceedings of the International Network Optimization Conference (INOC 2005)*, Lisbon, Portugal, March 2005. [cf. COST-279, TD(05)009]. 58, 59

[112] S. de Vuyst, S. Wittevrongel, and H. Bruneel. Analysis of a priority scheduling discipline with place reservation. In *Proceedings of the International Network Optimization Conference (INOC 2005)*, Lisbon, Portugal, March 2005. [cf. COST-279, TD(04)036]. 59, 60

[113] W. Burakowski and H. Tarasiuk. On new strategy for prioritising the selected flow in queueing system. Technical Report 257TD(00)03, COST-257, 2000. 59

[114] P. Gao, S. Wittevrongel, and H. Bruneel. Analysis of discrete-time buffers with geometric service times and muliple servers. In *High Performance Computing Symposium*, pages 294–299, San Diego, April 2002. [cf. COST-279, TD(02)035], [345]. 61, 224

[115] P. Gao, S. Wittevrongel, and H. Bruneel. Discrete-time multiserver buffer systems with correlated arrivals and geometric service times. In *The Conference on Design, Analysis, and Simulation of Distributed Systems*, pages 27–34, Orlando, April 2003. [cf. COST-279, TD(03)041]. 61

[116] P. Gao, S. Wittevrongel, and H. Bruneel. Delay against system contents in discrete-time G/Geom/c queue. *Electronics Letters*, 39(17):1290–1292, 2003. 61

[117] P. Gao, S. Wittevrongel, K. Laevens, and H. Bruneel. Dicrete-Time Multiserver Preemptive Resume Priority Queues with Geometric Service Times. Technical Report 279TD(04)019, COST-279, 2004. [cf. COST-279, TD(04)019]. 61

[118] D. Fiems, S. de Vuyst, and H. Bruneel. The combined gated-exhaustive vacation system in discrete time. *Performance Evaluation*, 49(1-4):227–239, 2002. [cf. COST-279, TD(03)017]. 61

[119] K. Tworus, S. de Vuyst, S. Wittevrongel, and H. Bruneel. Queueing Analysis of the Stop-and-Wait ARQ Protocol in a Wireless Environment. Technical Report 279TD(03)043, COST-279, 2003. [cf. COST-279, TD(03)043]. 62

[120] S. De Vuyst, S. Wittevrongel, and H. Bruneel. Delay analysis of the Stop-and-Wait ARQ protocol over a correlated error channel. In *Proceedings of HET-NETs 2004, Performance Modelling and Evaluation of Heterogeneous Networks, Ilkley, UK, July 26-28*, pages P21/1–P21/11, 2004. 63

[121] K. Spaey, T. Hofkens, and C. Blondia. Timescales in Models for Bursty Traffic. Technical Report 279TD(03)002, COST-279, 2003. [cf. COST-279, TD(03)002], [215]. 63, 105, 106, 211

[122] V. Ramaswami. Matrix analytic methods for stochastic fluid flows. In D. Smith and P. Hey, editors, *Proceedings of the 16th International Teletraffic Congress, Edinburgh, UK*, pages 1019–1030. Elsevier Science B.V., 1999. 64

[123] G. Latouche and V. Ramaswami. A logarithmic reduction algorithm for Quasi-Birth-and-Death processes. *Journal of Applied Probability*, 30:650–674, 1993. 64

[124] A. da Silva Soares and G. Latouche. Further results on the similarity between fluid queues and QBDs. In G. Latouche and P. Taylor, editors, *Matrix-Analytic Methods Theory and Applications—Proceedings of the 4th International Conference on Matrix-Analytic Methods*, pages 89–106. World Scientific, 2002. [cf. COST-279, TD(01)002]. 64

[125] D. Anick, D. Mitra, and M. M. Sondhi. Stochastic theory of a data-handling system with multiple sources. *The Bell System Technical Journal*, 61(8):1871–1894, 1982. 64, 68, 106

[126] N. Akar and K. Sohraby. Numerically stable solution of large-scale finite Fluid Queues. In *International Teletraffic Congress*, Berlin, Germany, September 2003. [cf. COST-279, TD(03)011],[346]. 65, 224

[127] M. Fiedler, P. Carlsson, and A. A. Nilsson. Voice and multi-fractal data traffic in the Internet. In *Proceedings of the 26th Annual IEEE Conference on Local Computer Networks (LCN 2001)*, pages 426–431, Tampa, USA, November 2001. [cf. COST-279, TD(01)008]. 65, 106, 107

[128] D. Zaragoza and C. A. Carvalho Belo. Queueing Behavior of Multiplexed General ON-OFF Sources: An Engineering Perspective. Technical Report 279TD(02)007, COST-279, 2002. [cf. COST-279, TD(02)007]. 65, 66

[129] M. Mandjes, D. Mitra, and W. Scheinhardt. Models of network access using Feedback Fluid Queues. *Queueing Systems*, 44:365–398, 2003. [cf. COST-279, TD(02)011], [347, 348]. 66, 224

[130] D. Sass, K. Dolzer, S. Bodamer, and M. Lorang. Analytic Fluid Flow Approach for Fair Queueing Systems. Technical Report 279TD(03)033, COST-279, 2003. [cf. COST-279, TD(03)033]. 67

[131] M. Fiedler and K. Tutschku. Application of the Stochastic Fluid Flow Model for bottleneck identification and classification. In *Proceedings of 2003 Design, Analysis, and Simulation of Distributed Systems (DASD 2003)*, pages 35–42, Orlando, USA, April 2003. [cf. COST-279, TD(02)046]. 68, 106, 111

[132] I. Norros. Most probable paths in Gaussian priority queues. Technical Report 257TD(99)16, COST-257, 1999. 69

[133] P. Mannersalo and I. Norros. A Most Probable Path approach to queueing systems with general Gaussian input. *Computer Networks*, 40:399–412, 2002. [cf. COST-279, TD(01)013]. 69, 71

[134] I. Norros. Most Probable Path techniques for Gaussian queueing systems. In *Networking 2002*, Pisa, Italy, 2002. [cf. COST-279, TD(02)003]. 69, 112

[135] M. Mandjes, P. Mannersalo, I. Norros, and M. van Uitert. Large Deviations of Infinite Intersections of Events in Gaussian Processes. Technical Report 279TD(04)015, COST-279, 2004. [cf. COST-279, TD(04)015]. 70, 71

[136] M. Mandjes, P. Mannersalo, and I. Norros. Large deviations of Gaussian tandem queues and resulting performance formulae. Technical Report PNA-E0412, ISSN 1386-3711, CWI, 2004. http://www.cwi.nl. 71

[137] G. Haßlinger and M. Fiedler. Network dimensioning for Gaussian traffic aggregated from Markovian ON-OFF sources. In *SPIE ITCOM 2002*, Boston, USA, July/August 2002. [cf. COST-279, TD(02)013]. 72

[138] T. Bonald and J. W. Roberts. Performance of bandwidth sharing mechanisms for service differentiation in the Internet. In *Proceedings ITC Specialists Seminar on IP Measurement, Modeling and Management, Monterey, USA*, pages 22/1–22/10, 2000. 72

[139] J. V. L. Beckers, I. Hendrawan, R. E. Kooij, and R.D.van der Mei. Generalized Processor Sharing models for Internet access lines. In *Proceedings 9th IFIP Conference on Performance Modeling and Evaluation of ATM & IP Networks, Budapest, Hungary*, pages 101–112, 2001. 72

[140] A. Riedl, T. Bauschert, M. Perske, and A. Probst. Investigation of the M/G/R Processor Sharing system for dimensioning IP access networks with elastic traffic. In *Proceedings 1st Polish-German Symposium of Telecomunication Systems*, 2000. 72

[141] S. F. Yashkov. Mathematical problems in the theory of Processor-Sharing queueing systems. *Journal of Soviet Mathematics*, 58:101–147, 1992. 73

[142] S. F. Yashkov. Processor-Sharing queues: some progress in analysis. *Queueing Systems*, 2:1–17, 1987. 73

[143] R. Litjens and R. J. Boucherie. Performance Analysis of Downlink Shared Channels in a UMTS Network. Technical Report 279TD(02)045, COST-279, 2002. [cf. COST-279, TD(02)045]. 73, 128

[144] R. Litjens, H. van den Berg, R. J. Boucherie, F. Roijers, and M. Fleuren. Performance analysis of Wireless LANs: an integrated packet/flow level approach. In *ITC-18*, Berlin, Germany, September 2003. [cf. COST-279, TD(03)001]. 73, 105, 133

[145] H. van den Berg, R. Litjens, and J. Laverman. HSDPA flow level performance: the impact of key system and environment aspects. In *MSWIM'04*, Venice, Italy, October 2004. [cf. COST-279, TD(04)021]. 73, 129

[146] R. Litjens, H. van den Berg, and M. J. Fleuren. Spatial Traffic Heterogeneity in HSDPA Networks and its Impact on Network Planning. Technical Report 279TD(05)002, COST-279, 2005. [cf. COST-279, TD(05)002]. 73, 130

[147] R. Litjens, H. van den Berg, and R. J. Boucherie. Throughput Measures for Processor Sharing Models. Technical Report 279TD(03)022, COST-279, 2003. [cf. COST-279, TD(03)022]. 74

[148] C. Douligeris. Multiobjective flow control in telecommunication networks. In *Proceedings of INFOCOM '92, Florence, Italy*, 1992. 74

[149] A. A. Kherani and A. Kumar. Performance analysis of TCP with non-persistent sessions. In *Proceedings of the Workshop on Modelling of Flow and Congestion Control, Paris, France*, 2000. 74

[150] A. A. Kherani and A. Kumar. Stochastic models for throughput analysis of randomly arriving elastic flows in the Internet. In *Proceedings of INFOCOM '02, New York, USA*, 2002. 74

[151] N. Benameur, S. Ben Fredj, F. Delcoigne, S. Oueslati-Boulahia, and J. W. Roberts. Integrated admission control for streaming and elastic

traffic. In *Proceedings of the 2nd International Workshop on Quality of Future Internet Services, Coimbra, Portugal*, 2001. 74

[152] T. Bonald and L. Massoulié. Impact of fairness on Internet performance. In *Proceedings of SIGMETRICS '01, Cambridge, USA*, 2001. 74

[153] L. Guo and I. Matta. Differentiated control of Web traffic: a numerical analysis. In *Proceedings of SPIE ITCOM'2002, Boston, USA*, 2002. 75

[154] H. Feng and V. Misra. Mixed scheduling disciplines for network flows. *ACM SIGMETRICS Performance Evaluation Review*, 31:36–39, 2003. 75

[155] K. Avrachenkov, U. Ayesta, P. Brown, and E. Nyberg. Differentiation between short and long TCP flows: predictability of the response time. In *Proceedings of IEEE Infocom 2004, Hong Kong*, pages 762–773, 2004. 75

[156] I. Rai, G. Urvoy-Keller, M. Vernon, and E. Biersack. Performance analysis of LAS-based scheduling disciplines in a packet switched network. In *Proceedings of ACM SIGMETRICS/PERFORMANCE 2004, New York, USA*, pages 106–117, 2004. 75

[157] S. Aalto, U. Ayesta, and E. Nyberg-Oksanen. Two-level Processor-Sharing scheduling disciplines: mean delay analysis. In *Proceedings of ACM SIGMETRICS/PERFORMANCE 2004, New York, USA*, pages 97–105, 2004. 75

[158] S. Aalto, U. Ayesta, and E. Nyberg-Oksanen. M/G/1/MLPS compared to M/G/1/PS. Technical Report RR-5219, INRIA, 2004. to appear in *Operations Research Letters*. 75

[159] S. Aalto and U. Ayesta. Mean Delay Comparison among Multilevel Processor Sharing Scheduling Disciplines. Technical Report 279TD(05)007, COST-279, 2005. [cf. COST-279, TD(05)007]. 75

[160] E. Kuumola, J. Resing, and J. Virtamo. Joint distribution of instantaneous and averaged queue length in an M/M/1/K system. In Phuoc Tran-Gia and J. Roberts, editors, *15th ITC Specialist Seminar on Internet Traffic Engineering and Traffic Management*, pages 58–67, Würzburg, Germany, July 2002. [cf. COST-279, TD(02)022]. 77

[161] U. Krieger. The BMAP/G/1 Queue with Feedback Operating in Synchronous Random Environment as a Model for a Telecommunication Channel with Performance Fluctuation. Technical Report

279TD(03)038, COST-279, 2003. [cf. COST-279, TD(03)038]. 78, 79

[162] A. W. Berger and W. Whitt. Effective bandwidths with priorities. *IEEE/ACM Transactions on Networking*, 6(4):447–460, 1998. 79

[163] O. Østerbø. An Approximative Method to Calculate the Distribution of End-to-End Delay in Packet Networks. Technical Report 279TD(02)033, COST-279, 2002. [cf. COST-279, TD(02)033]. 80

[164] D. De Vleeschauwer, G. H. Petit, B. Steyaert, S. Wittevrongel, and H. Bruneel. Calculation of end-to-end delay quantile in network of M/G/1 queues. *Electronics Letters*, 37:535–536, 2001. 80

[165] V. Inghelbrecht, B. Steyaert, S. Wittevrongel, and H. Bruneel. Analytic study of the interdeparture time characteristics in a multistage network. In *ITC 18*, volume 5b, pages 1191–1200, Berlin, Germany, August/September 2003. [cf. COST-279, TD(02)030]. 81, 103

[166] V. Inghelbrecht, B. Steyaert, and H. Bruneel. Study of the burstification mechanism of an OBS edge router. In *7th IFIP Working Conference on Optical Network Design & Modelling Conference (ONDM2003)*, volume 2, pages 1221–1239, Budapest, Hungary, February 2003. [cf. COST-279, TD(03)037]. 81, 161, 162

[167] N. Akar and E. Karasan. Exact calculation of blocking probabilities for bufferless optical burst switched links with partial wavelength conversion. In *Proceedings of 1st International Conference on Broadband Networks, BROADNETS*, 2004. [cf. COST-279, TD(04)031]. 82, 162, 163

[168] K. Laevens and H. Bruneel. Analysis of a single-wavelength optical buffer. In *INFOCOM 2003*, San Francisco, USA, April 2003. [cf. COST-279, TD(03)015]. 83, 166

[169] W. Rogiest, K. Laevens, D. Fiems, and H. Bruneel. Analysis of an Asynchronous Optical Buffer. Technical Report 279TD(05)012, COST-279, 2005. [cf. COST-279, TD(05)012]. 83, 84, 166

[170] J. Walraevens, S. Wittevrongel, and H. Bruneel. An analytic technique to evaluate the performance of optical packet switches. In *7th IFIP Working Conference on Optical Network Design & Modelling*, volume 2, pages 1171–1185, Budapest, Hungary, February 2003. [cf. COST-279, TD(02)036], [303]. 84, 166, 220

[171] D. Fiems, K. Laevens, and H. Bruneel. Performance analysis of an all-optical packet buffer. In *Proceedings of the 9th Conference on Optical Network Design & Modelling (ONDM 2005)*, Milano, Italy, February 2005. [cf. COST-279, TD(04)034]. 84, 166

[172] V. Paxson and S. Floyd. Wide-area traffic: The failure of Poisson modeling. *IEEE/ACM Transactions on Networking*, 3(3):226–244, June 1995. 85

[173] W. E. Leland, M. S. Taqqu, W. Willinger, and D. V. Wilson. On the self-similar nature of Ethernet traffic. In Deepinder P. Sidhu, editor, *ACM SIGCOMM*, pages 183–193, San Francisco, California, 1993. 85

[174] W. E. Leland, M. S. Taqqu, W. Willinger, and D. V. Wilson. On the self-similar nature of Ethernet traffic (extended version). *IEEE/ACM Transactions on Networking*, 2(1):1–15, February 1994. 85

[175] BC - Ethernet traces of LAN and WAN traffic. http://ita.ee.lbl.gov/html/contrib/BC.html, last checked March 06, 2005. 85

[176] K. Salamatian and S. Fdida. A Framework for Interpreting Measurement over Internet. Technical Report 279TD(04)009, COST-279, 2004. [cf. COST-279, TD(04)009]. 86, 108

[177] M. Franceschinis, M. Mellia, M. Meo, and M. Munafo. Measuring TCP over WiFi: A Real Case. Technical Report 279TD(04)038, COST-279, 2004. [cf. COST-279, TD(04)038]. 86, 88, 89, 97, 135

[178] L. Breuer, G. Latouche, and M.-A. Remiche. An EM Algorithm for the Model Fitting of a Class Transient Markovian Arrival Processes. Technical Report 279TD(04)023, COST-279, 2004. [cf. COST-279, TD(04)023]. 86, 101

[179] P. Carlsson, M. Fiedler, and A. Ekberg. On an Implementation of a Distributed Passive Measurement Infrastructure. Technical Report 279TD(03)042, COST-279, 2003. [cf. COST-279, TD(03)042], [349]. 86, 90, 224

[180] K. Salamatian, B. Baynat, and T. Bugnazet. Cross traffic estimation by loss process analysis. In *15th ITC Specialist Seminar on Internet Traffic Engineering and Traffic Management*, Würzburg, Germany, July 2002. [cf. COST-279, TD(02)015]. 87, 88, 90, 110

[181] P. Carlsson, M. Fiedler, K. Tutschku, S. Chevul, and A. A. Nilsson. Obtaining reliable bit rate measurements in SNMP-managed networks. In P. Tran-Gia and J. Roberts, editors, *Proceedings of 15th ITC Specialist Seminar on Internet Traffic Engineering and Traffic Management*, pages 114–123, Würzburg, Germany, July 2002. [cf. COST-279, TD(02)021]. 87

[182] J. Bachmann, M. Matthes, O. Drobnik, and U. Krieger. Mobility and QoS-management for adaptive applications. In *11th International World Wide Web Conference WWW2002*, Honolulu, USA, May 2002. [cf. COST-279, TD(02)001]. 88, 97, 139

[183] M. Fiedler, K. Tutschku, P. Carlsson, and A. A. Nilsson. Identification of performance degradation in IP networks using throughput statistics. In J. Charzinski, R. Lehnert, and P. Tran Gia, editors, *Providing Quality of Service in Heterogeneous Environments. Proceedings of the 18th International Teletraffic Congress (ITC-18)*, pages 399–407, Berlin, Germany, September 2003. [cf. COST-279, TD(03)021]. 88, 106, 111, 113

[184] TCPDUMP Public Repository. *Homepage*, 2004. http://www.tcpdump.org (Verified in January 2005). 88, 90

[185] J. Kilpi and I. Norros. Testing the Gaussian character of access network traffic. Technical Report 279TD(01)003, COST-279, 2001. [cf. COST-279, TD(01)003], [201]. 88, 96, 210

[186] A. Veres, Z. Kenesi, S. Molnár, and G. Vattay. On the propagation of long range dependence in the Internet. In *ACM SIGCOM 2000*, Stockholm, Sweden, August/September 2000. [cf. COST-279, TD(01)016], [200]. 88, 95, 210

[187] S. Molnár and G. Terdik. A general fractal model of Internet traffic. In *The 26th Annual IEEE Conference on Local Computer Networks*, Tampa, USA, November 2001. [cf. COST-279, TD(02)004]. 88, 89, 107

[188] J. Kilpi. A Portrait of a GPRS/GSM Session. Technical Report 279TD(02)040, COST-279, 2002. [cf. COST-279, TD(02)040], [202]. 88, 96, 97, 118, 210

[189] J. Kilpi. Distributional Properties of GPRS/GSM Session Volumes and Durations. Technical Report 279TD(03)023, COST-279, 2003. [cf. COST-279, TD(03)023]. 88, 97, 118

[190] L. Muscariello, M. Mellia, M. Meo, M. Ajmone Marsan, and R. Lo Cigno. An MMPP hierarchical model of Internet traffic. In *Proceedings of IEEE ICC 2004*, pages 353–364, June 2004. [cf. COST-279, TD(03)032]. 88, 89, 102

[191] K. Tutschku. A measurement-based traffic profile of the eDonkey file-sharing service. In *Proceedings of the 5th Passive and Active Measurement Workshop (PAM2004)*, pages 12–21, Antibes Juan-les-Pins, France, April 2004. [cf. COST-279, TD(03)049]. 88, 89, 99, 171

[192] D. Rossi, L. Muscariello, and M. Mellia. On the properties of TCP arrival process. In *Proceedings of IEEE ICC 2004*, pages 353–364, June 2004. [cf. COST-279, TD(04)013]. 88, 89, 95

[193] P. Arlos and M. Fiedler. A Comparison of Measurement Accuracy for DAG, `tcpdump` and `windump`. Technical Report 279TD(05)005, COST-279, 2005. [cf. COST-279, TD(05)005]. 88, 89, 90, 91

[194] WINDUMP. *Homepage*, 2004. http://windump.polito.it (Verified in January 2005). 88, 90

[195] A. de Vendictis, F. Vacirca, and A. Baiocchi. Experimental Analysis of TCP and UDP Traffic Performance over Infra-Structured 802.11b WLANs. Technical Report 279TD(04)033, COST-279, 2004. [cf. COST-279, TD(04)033]. 89, 90, 97, 134

[196] Meta Search Inc. eDonkey2000 Home Page. http://www.eDonkey2000.com/. 89, 99, 167, 171

[197] K. Tutschku and H. de Meer. A measurement study on signaling on Gnutella overlay networks. In *Fachtagung-Kommunikation in Verteilten Systemen (KiVS) 2003*, pages 295–306, Leipzig, Germany, February 2003. [cf. COST-279, TD(03)004]. 90, 97, 170

[198] Endace Measurement Systems. *Homepage*, 2004. http://www.endace.com (Verified in January 2005). 90

[199] G. Hu, K. Dolzer, and C. Gauger. Does burst assembly really reduce the self-similarity? In *Optical Fiber Communication 2003*, pages 124–126, Atlanta, Georgia, USA, March 2003. [cf. COST-279, TD(03)030]. 95, 159, 160

[200] A. Veres, Z. Kenesi, S. Molnár, and G. Vattay. TCP's role in the propagation of self-similarity in the Internet. *Computer Communications,*

Special Issue on Performance Evaluation of IP Networks, 26(8):899–913, May 2003. [cf. COST-279, TD(01)016], [186]. 95, 208

[201] J. Kilpi and I. Norros. Testing the Gaussian approximation of aggregate traffic. In *The 2nd Internet Measurement Workshop*, Marseille, France, 2002. [cf. COST-279, TD(01)003], [185]. 96, 208

[202] J. Kilpi. A portrait of a GPRS/GSM session. In *18th International Teletraffic Congress*, Berlin, Germany, August/September 2003. [cf. COST-279, TD(02)040], [188]. 96, 97, 118, 208

[203] T. Karagiannis, A. Broido, N. Brownlee, kc claffy, and M. Faloutsos. Is P2P dying or just hiding? In *Globecom 2004*, November/December 2004. http://www.caida.org/outreach/papers/2004/p2p-dying/. 97

[204] N. Markovitch and U. R. Krieger. The estimation of heavy-tailed probability density functions, their mixtures and quantiles. *Computer Networks*, 2002. [cf. COST-279, TD(01)001]. 99

[205] U. Krieger and N. Markovitch. On-Line Estimation of Heavy-Tailed Traffic Characteristics in Web Data Mining. Technical Report 279TD(03)039, COST-279, 2003. [cf. COST-279, TD(03)039]. 100

[206] G. Latouche and M.-A. Remiche. A MAP-based Poisson cluster model for Web traffic. *Performance Evaluation*, 49:359–370, 2002. [cf. COST-279, TD(02)016]. 100

[207] Z. Liu, N. Niclausse, and C. Jalpa-Villanueva. Traffic model and performance evaluation of Web servers. *Performance Evaluation*, 46:77–100, 2001. 100

[208] G. Latouche, M.-A. Remiche, and P. Taylor. Transient Markov arrival processes. *The Annals of Applied Probability*, 13:628–640, 2003. 100

[209] A. Andersen and B. Nielsen. A Markovian approach for modeling packet traffic with long-range dependence. *Journal on Selected Areas in Communications*, 16(5):719–732, 1998. 101

[210] A. Švigelj, M. Mohorčič, and G. Kandus. Traffic class dependent routing in packet-switched non-geostationary ISL networks. In *PIMRC 2002: IEEE International Symposium on Personal, Indoor and Mobile Radio Communications*, volume 3, pages 1382–1386, Lisbon, Portugal, September 2002. [cf. COST-279, TD(03)024], [350, 351, 352, 353, 354]. 101, 142, 224, 225

[211] P. Salvador, A. Pacheco, and R. Valadas. Multiscale fitting procedure using Markov Modulated Poisson Processes. *Telecommunication Systems Journal*, 23(1-2):123–148, June 2003. [cf. COST-279, TD(02)018]. 101

[212] A. Feldmann. Characteristics of TCP connection arrivals. In K. Park and W. Willinger, editors, *Self-similar Network Traffic and Performance Evaluation*. J. Wiley & Sons, 2000. 103

[213] P. Lassila, H. van den Berg, M. Mandjes, and R. Kooij. An integrated packet/flow model for TCP performance analysis. In *ITC 18*, pages 651–660, Berlin, Germany, August/September 2003. [cf. COST-279, TD(02)034]. 105

[214] J. Padhye, V. Firoiu, D. Towsley, and J. Kurose. Modeling TCP throughput: a simple model and its empirical validation. In *Proceedings of ACM SIGCOMM'98*, Vancouver, CA, September 1998. 105

[215] K. Spaey, D. de Vleeschauwer, T. Hofkens, and C. Blondia. Timescales in models for bursty traffic. In M. S. Obaidat, F. Davoli, E. Ferro, and I. Onyuksel, editors, *2003 International Symposium on Performance Evaluation of Computer and Telecommunication Systems (SPECTS 2003)*, volume 35(4) of *Simulation Series*, pages 465–473. SCS, July 2004. [cf. COST-279, TD(03)002], [121]. 106, 201

[216] P. Mannersalo, I. Norros, and R. Riedi. Multifractal products of stochastic processes: A preview. Technical Report 257TD(99)31, COST-257, 1999. 107

[217] P. Salvador, A. Nogueira, and R. Valadas. Joint characterization of the packet arrival and packet size processes of multifractal traffic based on Stochastic L-Systems. In *International Teletraffic Congress*, Berlin, Germany, September 2003. [cf. COST-279, TD(03)008]. 107

[218] E. Hyytiä, L. Nieminen, and J. Virtamo. Spatial Node Distribution in the Random Waypoint Mobility Model. Technical Report 279TD(04)029, COST-279, 2004. [cf. COST-279, TD(04)029]. 108, 144, 145

[219] E. Hyytiä and J. Virtamo. Random Waypoint model in n-dimensional space. *Operations Research Letters*, 2005. [cf. COST-279, TD(04)032]. 108, 144, 145

[220] A. Medina, N. Taft, K. Salamatian, S. Bhattacharyya, and C. Diot. Traffic matrix estimation: existing techniques and new directions. In

ACM SIGCOMM 2002, Pittsburg, USA, August 2002. [cf. COST-279, TD(02)014]. 110

[221] K. Salamatian, A. Soule, and N. Taft. Flow Classification by Histograms or How to Go to Safari over Internet. Technical Report 279TD(04)006, COST-279, 2004. [cf. COST-279, TD(04)006]. 111

[222] M.-I. Jeannin, L. Bernaille, A. Soule, and K. Salamatian. Blind Applicative Flow Recognition through Behavioral Classification. Technical Report 279TD(05)010, COST-279, 2005. [cf. COST-279, TD(05)010]. 112

[223] P. Mannersalo. Some notes on prediction of teletraffic. In *15th ITC Specialist Seminar*, pages 220–229, Würzburg, Germany, 2002. [cf. COST-279, TD(02)025]. 112

[224] 3GPP. The Third Generation Partnership Project, 1998. http://www.3gpp.org. 118

[225] F. Vacirca, T. Ziegler, and E. Hasenleithner. Large Scale Estimation of TCP Spurious Timeout Events in Operational GPRS Networks. Technical Report 279TD(05)003, COST-279, 2005. [cf. COST-279, TD(05)003]. 119

[226] M. Meo, M. Ajmone Marsan, and C. Batetta. Resource management policies in GPRS wireless Internet access systems. In *IEEE International Performance and Dependability Symposium*, Washington, USA, June 2002. [cf. COST-279, TD(02)028], [355]. 119, 225

[227] R. J. Boucherie and A. Ule. Adaptive dynamic channel borrowing in road-covering mobile networks. Technical Report 279TD(02)012, COST-279, 2002. [cf. COST-279, TD(02)012]. 120

[228] J. van Leeuwaarden, S. Aalto, and J. Virtamo. Load Balancing in Cellular Networks Using First Policy Iteration. Technical Report 279TD(02)023, COST-279, 2002. [cf. COST-279, TD(02)023]. 120

[229] D. Staehle and A. Mäder. An analytic approximation of the uplink capacity in a UMTS network with heterogeneous traffic. In *ITC-18*, Berlin, September 2003. [cf. COST-279, TD(03)018]. 123

[230] T. Hoßfeld and D. Staehle. Semi-Analytic Model of the UMTS Downlink Capacity with WWW Traffic on Dedicated Channels. Technical Report 279TD(04)026, COST-279, 2004. [cf. COST-279, TD(04)026]. 124

[231] S. Imre, P. Petrás, and R. Tancsics. Efficiency validation of 3G/4G WCDMA air interface call admission control in OMNeT++ environment. In *SoftCOM*, pages 852–858, Split, Dubrovnik (Croatia), Ancona, Venice (Italy), October 2003. FESB-Split, ISBN 953-6114-64-X. [cf. COST-279, TD(03)025], [356, 357]. 125, 225

[232] A. Irwan Endrayanto, H. van den Berg, and R. J. Boucherie. Characterizing CDMA Downlink Feasibility via Effective Interference. Technical Report 279TD(03)047, COST-279, 2003. [cf. COST-279, TD(03)047]. 125

[233] B. Zovko-Cihlar, W. Afric, and S. Grgic. Interference in Direct Spread Sequence Mobile Communication System. Technical Report 279TD(03)013, COST-279, 2003. [cf. COST-279, TD(03)013]. 125

[234] D. Staehle, K. Leibnitz, K. Heck, B. Schröder, A. Weller, and P. Tran-Gia. Analytical characterization of the soft handover gain in UMTS. In *The IEEE Vehicular Technology Conference*, Atlantic City, USA, October 2001. [cf. COST-279, TD(02)006]. 126

[235] K. Heck, D. Staehle, and K. Leibnitz. Diversity effects on the soft handover gain in UMTS networks. In *IEEE Vehicular Technology Conference*, Vancouver, Canada, 2002. [cf. COST-279, TD(02)042]. 127

[236] R. Litjens and H. van den Berg. Fair Adaptive Sheduling in Integrated Services UMTS Networks. Technical Report 279TD(02)016, COST-279, 2002. [cf. COST-279, TD(02)016]. 128, 129

[237] R. Litjens. The impact of mobility on UMTS network planning. *Computer Networks*, 38:497–515, 2002. [cf. COST-279, TD(01)011]. 131

[238] S. Szabo, S. Imre, and A. Burulitisz. On the accuracy of mobility modelling in wireless networks. In *John von Neumann Ph.D. Conference*, 2003. [cf. COST-279, TD(03)031], [358]. 131, 225

[239] IEEE. ANSI/IEEE Std 802.11, 1999 Edition, 1999. http://standards.ieee.org/getieee802. 133

[240] F. Roijers, X. Fan, and H. van den Berg. Analysis of TCP Flow Transfer Times over IEEE 802.11 Wireless LANs. Technical Report 279TD(04)024, COST-279, 2004. [cf. COST-279, TD(04)024]. 133

[241] J. L. Sobrinho, R. de Haan, and J. M. Brázio. Why RTS-CTS is not your ideal Wireless LAN multiple access protocol. In *Proceedings of*

IEEE WCNC 2005, New Orleans, USA, March 2005. [cf. COST-279, TD(04)035]. 134

[242] W. Burakowski, J. Sliwinski, and A. Beben. An Approach for Effective Handling of Constant Bit Rate Traffic over WLANs. Technical Report 279TD(04)028, COST-279, 2004. [cf. COST-279, TD(04)028]. 136

[243] J. Sliwinski, A. Beben, and W. Burakowski. Approach for Effective Handling of CBR Traffic over Wlans: Heterogeneous CBR Sources. Technical Report 279TD(05)013, COST-279, 2005. [cf. COST-279, TD(05)013]. 136

[244] M. Bottigliengo, C. Casetti, C.-F. Chiasserini, and M. Meo. Short-term fairness for TCP Flows in 802.1lb WLANs. In *IEEE Infocom*, Hong Kong, March 2004. [cf. COST-279, TD(04)022]. 137

[245] C. Blondia, O. Casals, L. Cerdà, N. van den Wijngaert, G. Willems, and P. de Cleyn. Low latency handoff mechanism and their implementation in an IEEE 802.11 network. In *18th ITC*, Berlin, Germany, September 2003. [cf. COST-279, TD(03)004]. 137

[246] C. Blondia, O. Casals, P. De Cleyn, and G. Willems. Performance analysis of IP micro-mobility handoff protocols. In *Proceedings of Protocols for High Speed Networks 2002 (PfHSN 2002)*, pages 211–226, Berlin, 2002. 138

[247] C. Blondia, O. Casals, Ll. Cerdà, and G.Willems. Performance analysis of a forwarding scheme for handoff in HAWAII. In *Proceedings of Networking 2002*, pages 504–514, Pisa, 2002. 138

[248] C. Blondia, O. Casals, N. Van den Wijngaert, and G. Willems. Performance analysis of smooth handoff in mobile IP. In *Proceedings of MSWiM'2002*, Atlanta, USA, September 2002. 138

[249] O. Casals, Ll. Cerdà, G. Willems, C. Blondia, and N. Van den Wijngaert. Performance evaluation of the Post-Registration method, a low latency handoff in MIPv4. In *Proceedings of ICC*, 2003. 138

[250] C. Blondia, O. Casals, Ll. Cerdà, N. Van den Wijngaert, G. Willems, and P. de Cleyn. Performance comparison of low latency Mobile IP schemes. In *Proceedings of WiOpt'03*, pages 115–124, INRIA Sophia Antipolis, France, March 2003. 138

[251] R. Schulcz, S. Szabó, S. Imre, and L. Pap. The effect of radio cell size in wmATM based Third Generation mobile systems. In *Services & Applications in the Wireless Public Infrastructure 2001*, pages 59–72, Paris, France, July 2001. HERMES Scientific Publications. [cf. COST-279, TD(01)015]. 140

[252] R. Pries and K. Heck. Performance Comparison of Handover Mechanisms in Wireless Lan Networks. Technical Report 279TD(04)025, COST-279, 2004. [cf. COST-279, TD(04)025]. 141

[253] L. Isaksson, M. Fiedler, and A. Nilsson. Validation of simulations of Bluetooth's Frequency Hopping Spread Spectrum technique. In *2004 Design, Analysis, and Simulation of Distributed Systems (DASD 2004)*, Arlington, Virginia, April 2004. [cf. COST-279, TD(04)002]. 143

[254] L. Isaksson and M. Fiedler. Optimization of the random backoff boundary of the Bluetooth Frequency Hopping Spread Spectrum Technique. In *2005 Design, Analysis, and Simulation of Distributed Systems (DASD 2005)*, San Diego, CA, April 2005. [cf. COST-279, TD(04)027]. 144

[255] D. B. Johnson and D. A. Maltz. Dynamic source routing in ad hoc wireless networks. In Imielinski and Korth, editors, *Mobile Computing*, volume 353. Kluwer Academic Publishers, 1996. 145

[256] M. Grossglauser and D. N. C. Tse. Mobility increases the capacity of ad hoc wireless networks. *IEEE/ACM Transactions on Networking*, 10(4), August 2002. 145

[257] P. Gupta and P. R. Kumar. The capacity of wireless networks. *IEEE Transactions on Information Theory*, 46(2), March 2000. 145

[258] O. Dousse, P. Thiran, and M. Hasler. Connectivity in ad-hoc and hybrid networks. In *Proc of the INFOCOM*, New York, June 2002. 145

[259] M. D. Penrose. On k-connectivity for a geometric random graph. *Random Structures and Algorithms*, 15(2):145–164, 1999. 145

[260] C. Bettstetter and C. Wagner. The spatial node distribution of the Random Waypoint mobility model. In *Proceedings of German Workshop on Mobile Ad Hoc networks (WMAN)*, Ulm, Germany, March 2002. 145

[261] C. Bettstetter, G. Resta, and P. Santi. The node distribution of the Random Waypoint mobility model for wireless ad hoc networks. *IEEE Transactions on Mobile Computing*, 2(1):25–39, 2003. 145

[262] C. Bettstetter, H. Hartenstein, and X. Pérez-Costa. Stochastic properties of the Random Waypoint mobility model. *ACM/Kluwer Wireless Networks: Special Issue on Modeling and Analysis of Mobile Networks*, 10(5), September 2004. 145

[263] W. Navidi and T. Camp. Stationary distributions for the Random Waypoint mobility model. *IEEE Transactions on Mobile Computing*, 3(1):99–108, January-March 2004. 145

[264] J. Le Boudec. On the stationary distribution of speed and location of Random Waypoint. *IEEE Transactions on Mobile Computing*, 2004. 145

[265] Esa Hyytiä and Jorma Virtamo. Random Waypoint model in n-dimensional space. *Operations Research Letters*, to appear in, 2005. 145

[266] S. Imre and M. Szalay. LTRACK—A Novel Location Management Method. Technical Report 279TD(04)001, COST-279, 2004. [cf. COST-279, TD(04)001]. 146

[267] M. Szalay and S. Imre. Reliability modelling of tree topology IP micro mobility networks. In *EUNICE 2002 7th Open European Summer School on Adaptable Networks and Teleservices*, pages 63–69, Trondheim, Norway, September 2002. Tapir Uttrykk. [cf. COST-279, TD(03)009], [359, 360]. 146, 225

[268] A. de Vendictis, F. Vacirca, and A. Baiocchi. TCP over wireless links with Rayleigh fading: some considerations about channel energy efficiency and TCP performance. In *16th ITC Specialist Seminar*, Antwerp, Belgium, August/September 2004. [cf. COST-279, TD(04)007]. 147

[269] R. Khalili and K. Salamatian. On the Distribution of Error Events Length in Convolutional Codes. Technical Report 279TD(04)008, COST-279, 2004. [cf. COST-279, TD(04)008]. 147

[270] G. Falcao and S. Michandani. Extending optical reach and transparency—Industry trend or event. *Telecommunications*, 2000. 151

[271] B. Ramamurthy, S. Yaragorla, and X. Yang. Translucent optical WDM networks for the next generation backbone networks. In *GLOBECOM'2001*, New York, USA, 2001. 151

[272] G. Shen and W. D. Grover. Segment-based approaches to survivable translucent network design under various ultra-long-haul system reach capabilities. *Journal of Optical Networking*, 3(1), January 2004. 151

[273] E. Yetginer and E. Karasan. Regenerator placement and traffic engineering with restoration in GMPLS networks. *Photonic Network Communications*, 6(2):139–149, September 2003. 151

[274] M. Arisoylu and E. Karasan. Design of translucent optical networks: partitioning and restoration. *Photonic Network Communications*, 8:209–221, 2004. [cf. COST-279, TD(02)009], [361, 362]. 152, 155, 226

[275] E. Karasan and E. Goldstein. Optical restoration at the wavelength-multiplex-section level in WDM mesh networks. *IEEE Photonic Technology Letters*, 10(9):1343–1345, Sep 1998. 152

[276] R. Ramaswami and K. N. Sivarajan. Routing and wavelength assignment in all-optical networks. *IEEE/ACM Transactions on Networking*, 3(5):489–500, 1995. 152

[277] L. Li and A. K. Somani. Dynamic wavelength routing using congestion and neighborhood information. *IEEE/ACM Transactions on Networking*, 7(5):779–786, 1999. 152

[278] D. Benerjee and B. Mukherjee. Wavelength-routed optical networks: Linear formulation, resource budgeting tradeoffs and a reconfigurability study. *IEEE-ACM Transactions on Networking*, 8(5):598–606, 2000. 152

[279] E. Karasan and E. Ayanoglu. Effects of wavelength routing and selection algorithms on wavelength conversion gain in WDM optical networks. *IEEE/Transactions on Networking*, 6(2):186–196, April 1998. 152

[280] K. Zhu and B. Mukherjee. Traffic grooming in an optical WDM mesh network. *IEEE Journal on Selected Areas in Communications*, 20(1), Jan 2002. 152

[281] R. Dutta and G. N. Rouskas. Traffic grooming in WDM networks: Past and future. *IEEE Network Mag.*, 16(6), November/December 2002. 152

[282] L. Noirie, M. Vigoureux, and E. Dotaro. Impact of intermediate grouping on the dimensioning of multi-granularity optical networks. In *Optical Fiber Conference*, Anaheim, USA, 2001. 152, 153

[283] H. Zhu, *et al.* Cost-effective WDM backbone network design with OXCs of different bandwidth granularities. *IEEE Journal on Selected Areas in Communications*, 21(9), November 2003. 152

[284] X. Cao, V. Anand, Y. Xiong, and C. Qiao. A study of waveband switching with multi-layer multi-granular optical cross-connects. *IEEE Journal on Selected Areas in Communications*, 21(7):1081–1095, Sepember 2003. 152

[285] X. Cao, V. Anand, and C. Qiao. Multi-layer versus single-layer Optical Cross-connect architectures for waveband switching. In *INFOCOM*, Hong Kong, 2004. 152

[286] P. H. Ho and H. T. Mouftah. Routing and wavelength assignment with multi-granularity traffic in optical networks. *Journal of Lightwave Technology*, 20(8):1292–1303, August 2002. 154

[287] S. Subramaniam, M. Azizoglu, and A. K. Somani. On optimal converter placement in wavelength-routed networks. *IEEE/ACM Transactions on Networking*, 7(5):754–766, 1999. 154

[288] E. Karasan, O. E. Karasan, and G. Erdogan. Optimum placement of wavelength interchanging nodes in optical networks with sparse conversion. In *9th European Conference on Networks & Optical Communications*, Eindhoven, Netherlands, 2004. 154, 155, 156

[289] N. Sengezer and E. Karasan. TSCP: A tabu search algorithm for wavelength converting node placement in WDM optical networks. In A. Pattavina, editor, *Optical Network Design and Modelling Conference*, pages 359–369, Milan, Italy, February 2005. [cf. COST-279, TD(05)019],[363]. 154, 155, 156, 226

[290] B. Puype, *et al.* Multi-layer traffic engineering in data-centric optical networks. In *COST266/IST Optimist Workshop at Optical Network Design and Modelling Conference*, Budapest, Hungary, 2003. 154

[291] B. Puype, *et al.* Optical cost metrics in multi-layer traffic engineering for IP-over-Optical networks. In *6th International Conference on Transparent Optical Networks*, Wroclaw, Poland, 2004. 154

[292] C. Ou, *et al.* Sub-path protection for scalability and fast recovery in optical WDM mesh networks. *IEEE Journal on Selected Areas in Communications*, 22:1859–1875, November 2004. 155

[293] E. Karasan, O. Karasan, N. Akar, and M. Pinar. Mesh topology design in overlay Virtual Private Networks. *Electronics Letters*, 38(16):939–941, 2002. [cf. COST-279, TD(03)010], [364]. 156, 226

[294] M. Yoo and C. Qiao. Optical Burst Switching (OBS)—A new paradigm for an optical Internet. *Journal of High-Speed Networks*, 8(1):69–84, 1999. 158

[295] C. Gauger and M. Köhn and J. Scharf. Comparison of contention resolution strategies in OBS network scenarios. In *6th International Conference on Transparent Optical Networks*, Wroclaw, Poland, 2004. [cf. COST-279, TD(01)009], [300]. 158, 219

[296] J. Xu and *et al.* Efficient channel scheduling algorithms in optical burst switched networks. In *INFOCOM*, San Francisco, USA, 2003. 158

[297] Y. Xiong, M. Vanderhoute, and H. C. Cankaya. Control architecture in optical burst-switched WDM networks. *IEEE Journal on Selected Areas Communications*, 18(10):1838–1851, October 2000. 159

[298] M. Yoo, C. Qiao, and S. Dixit. QoS performance of optical burst switching in IP-over-WDM networks. *IEEE Journal on Selected Areas in Communications*, 18(10):2062–71, October 2000. 159

[299] Y. Chen, C. Qiao, and X. Yu. Optical Burst Switching: A new area in optical networking research. *IEEE Network Magazine*, 18(3):16–23, May/June 2004. 159

[300] K. Dolzer and C. Gauger. On burst assembly in Optical Burst Switching networks—a performance evaluation of Just-Enough-Time. In *17th International Teletraffic Congress*, pages 149–160, Salvador da Bahia, Brazil, November 2001. [cf. COST-279, TD(01)009], [295]. 162, 163, 219

[301] J. White, M. Zukerman, and H. L. Vu. A framework for Optical Burst Switching network design. *IEEE Communications Letters*, 6:268–270, June 2002. 162

[302] L. Nieminen and E. Hyytiä. Delay line configurations in Optical Burst Switching with JET protocol. In *16th Nordic Teletraffic Seminar*, pages 229–240, Espoo, Finland, 2002. [cf. COST-279, TD(02)024]. 165

[303] J. Walraevens, S. Wittevrongel, and H. Bruneel. Calculation of the packet loss in optical packet switches: an analytic technique. *AEU*

(International Journal of Electronics and Communications), 57(4):270–276, 2003. [cf. COST-279, TD(02)036], [170]. 166, 206

[304] Anonymous. The Gnutella Protocol Specification v0.4. Available at http://dss.clip2.com, Clip2 Distributed Search Solutions, 2001. 167

[305] Sharman Networks. KaZaA Media Desktop. http://www.kazaa.com/. 167

[306] Internet2 project. NetFlow Statistics Home Page. http://netflow.internet2.edu/weekly/. 167

[307] Sprint Advanced Technology Lab. IP Monitoring Project (IPMON) Home Page. http://ipmon.sprintlabs.com/ipmon.php/. 167

[308] S. Sen and J. Wang. Analyzing Peer-to-Peer traffic across large networks. *ACM/IEEE Transactions on Networking*, 12(2), April 2004. 167

[309] N. Ben Azzouna and F. Guillemin. Analysis of ADSL traffic on an IP backbone link. In *GLOBECOM 2003*, San Francisco, California, December 2003. 167

[310] J. Enríquez Gabeiras. Panel Presentation on "Issues in Peer-to-Peer Networking" at COST279 Mid-Seminar, University of Roma "La Sapienza", January 21–22, 2004, Italy. 167

[311] D. Barkai. *Peer-to-Peer Computing*. Intel Press, Hillsborow, OR, 2001. 168

[312] H. Balakrishnan, M. F. Kaashoek, D. Karger, R. Morris, and I. Stoica. Looking up data in P2P systems. *Communications of the ACM*, 43(2), February 2003. 169

[313] Skype Technologies S.A. Skype homepage. http://www.skype.com/. 169

[314] T. Hoßfeld, K. Tutschku, and F.-U. Andersen. Mapping of file-sharing onto mobile environments: enhancement by UMTS. In *Mobile Peer-to-Peer Computing MP2P, in Conjunction with the 3rd IEEE International Conference on Pervasive Computing and Communications (PerCom'05)*, Kauai Island, Hawaii, March 2005. [cf. COST-279, TD(05)008]. 171

[315] J. Oberender, F.-U. Andersen, H. de Meer, I. Dedinski, T. Hoßfeld, C. Kappler, A. Mäder, and K. Tutschku. Enabling mobile Peer-to-Peer networking. In *Mobile and Wireless Systems, LNCS 3427*, Dagstuhl, Germany, January 2005. 172

[316] G. Wearden. eDonkey Pulls Ahead in Europe P2P Race. http://news.com.com/2100-1025-5091230.html. 172

[317] T. Hoßfeld, K. Tutschku, F.-U. Andersen, H. de Meer, and J. Oberender. Simulative performance evaluation of a mobile Peer-to-Peer file-sharing system. In *Next Generation Internet Networks NGI2005*, Rome, Italy, April 2005. 172

[318] A. Binzenhöfer and P. Tran-Gia. Delay analysis of a Chord-based Peer-to-Peer file-sharing system. In *ATNAC 2004*, Sydney, December 2004. [cf. COST-279, TD(04)010]. 173

[319] I. Norros and H. Reittu. On the Performance and Stability of Peer-to-Peer Networks with Highly Variable Life Times of Nodes. Technical Report 279TD(05)017, COST-279, 2005. [cf. COST-279, TD(05)017]. 174

[320] S. Saroiu, K. Gummadi, and S. Gribble. Measuring and analyzing the characteristics of Napster and Gnutella hosts. *Multimedia Systems*, 9(2), August 2003. 174

[321] M. Luisa Garcia Osma, F.-J. Ramón Salguero, G. Garcia de Blas, J. Andrés Colás, and J. Enríquez Gabeiras. Enabling Local Preference in Peer-to-Peer Traffic. Technical Report 279TD(04)017, COST-279, 2004. [cf. COST-279, TD(04)017]. 174

[322] N. Leibowitz, A. Bergman, R. Ben-Shaul, and A. Shavit. Are file swapping networks cacheable? Characterizing P2P traffic. In *7th International Workshop on Web Content Caching and Distribution (WCW'03)*, Boulder, USA, August 2002. 175

[323] K. P. Gummadi, R. J. Dunn, S. Saroiu, S. D. Gribble, H. M. Levy, and J. Zahorjan. Measurement, modeling, and analysis of a Peer-to-Peer file-sharing workload. In *Proceedings of the 19th ACM Symposium on Operating Systems Principles (SOSP-19)*, Bolton Landing (Lake George), USA, October 2003. 175

[324] J. Chu, K. Labonte, and B. N. Levine. Availability and locality measurements of Peer-to-Peer file sharing systems. In *in Proc. of SPIE ITCom:*

Scalability and Traffic Control in IP Networks, vol. 4868, Boston, USA, July/August 2002. 175

[325] eMule Project. Help Page on "Rating and Score". Available at http://www.emule-project.net/. 175

[326] WebHosting.Info Project. The IP to Country Database Home Page. http://ip-to-country.webhosting.info/. 175

[327] J. Luo, P. Th. Eugster, and J.-P. Hubaux. Route Driven Gossip: Probabilistic reliable multicast in ad hoc networks. In *INFOCOM 2003*, pages 2229–2239, San Francisco, California, March/April 2003. 176

[328] R. Chandra, V. Ramasubramanian, and K. Birman. Anonymous gossip: Improving multicast reliability in mobile ad-hoc networks. In *Proc. 21st International Conference on Distributed Computing Systems (ICDCS)*, pages 275–283, Phoenix (Mesa), Arizona, USA, April 2001. 176

[329] J. Luo amd P. Th. Eugster and J.-P. Hubaux. Pilot: Probabilistic lightweight group communication system for ad hoc networks. *IEEE Transactions on Mobile Computing*, 3(2), April/June 2004. 176

[330] P.T. Eugster, R. Guerraoui, A.-M. Kermarrec, and L. Massoulie. Epidemic information dissemination in distributed systems. *IEEE Computer*, 37(5):60–67, May 2004. 177

[331] A. Alagoz, E. Ahi, and Ö. Özkasap. Network Awareness and Buffer Management in Epidemic Information Dissemination. Technical Report 279TD(05)015, COST-279, 2005. [cf. COST-279, TD(05)015]. 177

[332] S. Aalto and E. Nyberg. Flow level models of DiffServ packet level mechanisms. In *Proceedings of the Sixteenth Nordic Teletraffic Seminar, NTS 16*, pages 194–205, Espoo, Finland, August 2002. [cf. COST-279, TD(01)004], [7, 333]. 190, 222

[333] E. Nyberg and S. Aalto. Differentiation and interaction of traffic: a flow level study. In *Proceedings of International Workshop, Art-Qos 2003*, pages 276–290, Warsaw, Poland, March 2003. [cf. COST-279, TD(01)004], [7, 332]. 190, 222

[334] H. Tuan Tran and T. Ziegler. An admission control scheme for voice traffic over IP networks. In *LNCS 2720, Proceedings of the 6th IEEE International Conference on High Speed Networks and Multimedia Communications HSNMC'03*, pages 353–364, July 2003. [cf. COST-279, TD(02)037], [16]. 191

[335] M. Menth, J. Milbrandt, and A. Reifert. Sensitivity of backup capacity requirements to traffic distribution and resilience constraints. In 1st *Conference on Next Generation Internet Networks Traffic Engineering (NGI)*, Rome, Italy, April 2005. [cf. COST-279, TD(03)046], [41]. 193

[336] T. Bonald, P. Olivier, and J. Roberts. Dimensioning IP access links carrying data traffic. *Annals of Telecommunications, Special Issue on Traffic Engineering and Routing*, 59(11-12), 2004. [cf. COST-279, TD(03)007], [54]. 194

[337] F. Delcoigne, A. Proutière, and G. Régnié. Modeling integration of streaming and data traffic. *Performance Evaluation*, 55(3-4):185–209, 2004. [cf. COST-279, TD(02)019], [62]. 195

[338] M. Klimo. Cell loss noise in the case of linear reconstruction. Porto Carras, Greece, June 1998. [cf. COST-279, TD(05)001], [339, 340]. 197, 223

[339] T. Kováčiková, M. Klimo, and P. Segeč. *Voice over Internet Protokol.* ES, Žilina, Slovakia, 2005. [cf. COST-279, TD(05)001], [338, 340]. 197, 223

[340] M. Klimo. Voice over IP: packet loss and jitter in E-model. November 1999. [cf. COST-279, TD(05)001], [338, 339]. 197, 223

[341] M. Ajmone Marsan, M. Franceschinis, E. Leonardi, F. Neri, and A. Tarello. Underload instabilities in packet networks with flow schedulers. *IEEE/ACM Transactions on Networking*, December 2004. [cf. COST-279, TD(02)027], [82]. 197

[342] O. Alparslan, N. Akar, and E. Karasan. AIMD-based online MPLS traffic engineering for TCP flows via distributed multi-path routing. *Annales Des Telecommunications-Annals of Telecommunications*, November/December 2004. [cf. COST-279, TD(04)020],[91]. 198

[343] M. Çaglar and Ö. Özkasap. Multicast transport protocol analysis: self-similar sources. In *4th Int. IFIP-TC6 Networking Conference*, Athens, Greece, May 2004. [cf. COST-279, TD(03)036], [94]. 199

[344] D. Fiems, B. Steyaert, and H. Bruneel. Discrete-time queues with generally distributed service times and renewal-type interruptions. *Performance Evaluation*, 55(3-4):277–298, 2004. [cf. COST-279, TD(01)005], [109]. 200

[345] P. Gao, S. Wittevrongel, and H. Bruneel. Discrete-time multiserver queues with geometric service times. *Computers & Operations Research*, 31(1):81–99, 2004. [cf. COST-279, TD(02)035], [114]. 201

[346] N. Akar and K. Sohraby. Infinite/finite buffer Markov Fluid Queues: a unified analysis. *Journal of Applied Probability*, 2004. [cf. COST-279, TD(03)011],[126]. 202

[347] M. Mandjes, D. Mitra, and W. Scheinhardt. A simple model of network access: feedback adaptation of rates and admission control. *Computer Networks*, 41:489–504, 2003. [cf. COST-279, TD(02)011], [129, 348]. 202, 224

[348] M. Mandjes, D. Mitra, and W. Scheinhardt. A simple model of network access: feedback adaptation of rates and admission control. In *Proceedings INFOCOM 2002*, pages 3–12, New York, US, July 2002. [cf. COST-279, TD(02)011], [129, 347]. 202, 224

[349] P. Carlsson, M. Fiedler, and A. Nilsson. A distributed passive measurement infrastructure. In *Proceedings of the 6th International Workshop on Passive and Active Network Measurement (PAM 2005)*, pages 215–227, Boston, MA, USA, March/April 2005. Springer. [cf. COST-279, TD(03)042], [179]. 207

[350] A. Švigelj, M. Mohorčič, and G. Kandus. Adaptive packet routing based on traffic class differentiation in Intersatellite Link networks. *WSEAS Transactions on Communications*, 1:138–143, 2002. [cf. COST-279, TD(03)024], [210, 351, 352, 353, 354]. 210, 224, 225

[351] A. Švigelj, M. Mohorčič, and G. Kandus. Traffic class dependent routing in packet-switched non-geostationary ISL networks. In E. del Re, editor, *Mobile and Personal Satellite Communications 5: Proceedings of the Fifth European Workshop on Mobile/Personal Satcoms (EMPS 2002)*, pages 45–52, Baveno, Italy, September 2002. [cf. COST-279, TD(03)024], [210, 350, 352, 353, 354]. 210, 224, 225

[352] A. Švigelj, M. Mohorčič, and G. Kandus. Traffic class dependent routing in ISL network with adaptive forwarding based on local link load information. In *Satellite Communications—From Fade Mitigation to Service Provision: International Workshop of COST Actions 272 and 280*, pages 395–402, Noordwijk, The Netherlands, May 2003. COST-272 / COST-280, ESTEC/ESA. [cf. COST-279, TD(03)024], [210, 350, 351, 353, 354]. 210, 224, 225

[353] G. Kandus, A. Švigelj, and M. Mohorčič. The impact of different scheduling policies on traffic class dependent routing in Intersatellite Link network. In *International Conference on Advanced Satellite Mobile Systems (ASMS 2003)*, ESA SP-541, Frascati, Italy, July 2003. ESA/ESTEC. [cf. COST-279, TD(03)024], [210, 350, 351, 352, 354]. 210, 224, 225

[354] A. Švigelj, M. Mohorčič, G. Kandus, A. Kos, M. Pustišek, and J. Bešter. Routing in ISL networks considering empirical IP traffic. *IEEE Journal on Selected Areas in Communications*, February 2004. [cf. COST-279, TD(03)024], [210, 350, 351, 352, 353]. 210, 224, 225

[355] M. Meo and M. Ajmone Marsan. Resource management policies in GPRS systems. *Perfomance Evaluation*, 2002. [cf. COST-279, TD(02)028], [226]. 212

[356] S. Imre. Dynamically optimised Chernoff bound based CAC for 3G/4G WCDMA systems. In *Microcoll*, pages 27–30, Budapest, Hungary, September 2003. Microcoll, ISBN 963-212-166-X. [cf. COST-279, TD(03)025], [231, 357]. 213, 225

[357] S. Imre, K. Hankó, P. Petrás, and R. Tancsics. Efficient call admission control method for 3G/4G WCDMA networks. In *CONTEL*, pages 293–300, Zagreb, Croatia, June 2003. CIP Zagreb, ISBN 953-184-055-5. [cf. COST-279, TD(03)025], [231, 356]. 213, 225

[358] S. Szabo, S. Imre, and A. Burulitisz. On the accuracy of mobility modelling in wireless networks. In *IEEE International Conference on Communications, ICC2004*, pages 20–24, Paris, France, June 2004. ISBN 0-7803-8534-9. [cf. COST-279, TD(03)031], [238]. 213

[359] M. Szalay and S. Imre. Reliability considerations of IP micro mobility networks. In *Design of Reliable Communication Networks DRCN*, pages 72–77, Budapest, Hungary, October 2001. [cf. COST-279, TD(03)009], [267, 360]. 216, 225

[360] M. Szalay and S. Imre. Parameter considerations of LTRAC. In *EUNICE 2004 10th Open European Summer School on Advances in Fixed and Mobile Networks*, pages 88–91, Tampere, Finland, June 2004. ISBN 952-15-1187-7. [cf. COST-279, TD(03)009], [267, 359]. 216, 225

[361] M. Arisoylu and E. Karasan. Subnetwork partitioning in translucent optical networks. In *The 6th Symposium on Computer Networks*, Cyprus, June 2001. [cf. COST-279, TD(02)009], [362, 274]. 217, 226

[362] M. Arisoylu and E. Karasan. Subnetwork partitioning and section restoration in translucent optical networks. In *Opticomm 2003: Optical Networking and Communications*, Dallas, USA, October 2003. [cf. COST-279, TD(02)009], [361, 274]. 217, 226

[363] N. Sengezer and E. Karasan. A tabu search algorithm for sparse placement of wavelength converters in optical networks. In C. Aykanat, T. Dayar, and I. Korpeoglu, editors, *ISCIS*, number 3280 in *Lecture Notes in Computer Science*, pages 247–256, Antalya, Turkey, 2004. Springer. [cf. COST-279, TD(05)019],[289]. 218

[364] E. Karasan, O. Karasan, N. Akar, and M. Pinar. Topology design in Virtual Private Networks. In *INFORMS National Meeting*, San Jose, USA, November 2002. [cf. COST-279, TD(03)010], [293]. 219

List of Temporary Documents

[TD(01)001] U. Krieger and N. M. Markovich. The Estimation of Heavy-Tailed Probability Density Functions. Technical Report 279TD (01)001, COST-279, 2001.

[TD(01)002] A. da Silva Soares and G. Latouche. Algorithmic Approach to Fluid Queues. Technical Report 279TD(01)002, COST-279, 2001.

[TD(01)003] J. Kilpi and I. Norros. Testing the Gaussian Character of Access Network Traffic. Technical Report 279TD(01)003, COST-279, 2001.

[TD(01)004] E. Nyberg, J. Virtamo, and S. Aalto. Relating Flow Level Requirements to DiffServ Packet Level Mechanisms. Technical Report 279TD(01)004, COST-279, 2001.

[TD(01)005] D. Fiems, H. Bruneel, and B. Steyaert. Analysis of a Discrete-Time Queueing Model with Server Interruptions Modeling Preemptive Priority Systems. Technical Report 279TD(01)005, COST-279, 2001.

[TD(01)006] J. Walraevens, B. Steyaert, and H. Bruneel. A Single-Server Queue with a Priority Scheduling Discipline: Performance Study. Technical Report 279TD(01)006, COST-279, 2001.

[TD(01)007] A. Jena and A. Popescu. Traffic Engineering for Internet Services. Technical Report 279TD(01)007, COST-279, 2001.

[TD(01)008] M. Fiedler, P. Carlsson, and A. Nilsson. Fluid Flow Analysis for Voice over IP and Multi-Fractal Data Traffic. Technical Report 279TD(01)008, COST-279, 2001.

[TD(01)009] K. Dolzer and C. Gauger. On Burst Assembly in Optical Burst Switching Networks—A Performance Evaluation of Just Enough-Time. Technical Report 279TD(01)009, COST-279, 2001.

[TD(01)010] M. Menth. A Scalable Protocol Architecture for End-to-End Signaling and Resource Reservation in IP Networks. Technical Report 279TD(01)010, COST-279, 2001.

[TD(01)011] R. Litjens. The Impact of Mobility on UMTS Network Planning. Technical Report 279TD(01)011, COST-279, 2001.

[TD(01)012] R. van der Mei, H. van den Berg, R. Vranken, and B. Gijsen. Analysis of a Flow Level Model for TCP Behavior in Case of Priority Queueing. Technical Report 279TD(01)012, COST-279, 2001.

[TD(01)013] I. Norros and P. Mannersalo. A Most Probable Path Approach to Queueing Systems with General Gaussian Input. Technical Report 279TD(01)013, COST-279, 2001.

[TD(01)014] P. Olivier and N. Benameur. Flow Level IP Traffic Characterization. Technical Report 279TD(01)014, COST-279, 2001.

[TD(01)015] R. Schulcz and S. Imre. Handover Support in WATM Networks. Technical Report 279TD(01)015, COST-279, 2001.

[TD(01)016] S. Molnár, Z. Kenesi, A. Veres, and G. Vattay. Self-Similarity Propagated over the Internet. Technical Report 279TD(01)016, COST-279, 2001.

[TD(01)017] W. Burakowski, A. Bak, F. Ricciato, S. Salsano, and H. Tarasiuk. Traffic Handling in AQUILA QoS IP Network. Technical Report 279TD(01)017, COST-279, 2001.

[TD(02)001] J. Bachmann, M. Matthes, O. Drobnik, and U. Krieger. Traffic Mobility in Wireless IP-Networks Subject to Mobility- and Resource-Management Conditions. Technical Report 279TD (02)001, COST-279, 2002.

[TD(02)002] H. Reittu and I. Norros. On Large Random Graphs of the "Internet Type". Technical Report 279TD(02)002, COST-279, 2002.

[TD(02)003] I. Norros. On a Gaussian Queue with Bandwith Allocation by Prediction. Technical Report 279TD(02)003, COST-279, 2002.

[TD(02)004] S. Molnár and G. Terdik. A Monofractal Model for Network Traffic. Technical Report 279TD(02)004, COST-279, 2002.

[TD(02)005] J. Milbrandt, D. Staehle, S. Köhler, and L. Berry. Decomposition of Large IP Networks for Routing Optimization. Technical Report 279TD(02)005, COST-279, 2002.

[TD(02)006] D. Staehle, K. Leibnitz, K. Heck, P. Tran-Gia, B. Schröder, and A. Weller. Analytical Characterization of the Soft Handover Gain in UMTS. Technical Report 279TD(02)006, COST-279, 2002.

[TD(02)007] D. Zaragoza and C. C. Belo. Queueing Behavior of Multiplexed General ON-OFF Sources: An Engineering Perspective. Technical Report 279TD(02)007, COST-279, 2002.

[TD(02)008] R. Boël and S. de Vuyst. Prediction Based Resource Allocation, a Simulation Experiment. Technical Report 279TD(02)008, COST-279, 2002.

[TD(02)009] E. Karasan and M. Arisoylu. Subnetwork Partitioning and Section Restoration in Translucent Optical Networks. Technical Report 279TD(02)009, COST-279, 2002.

[TD(02)010] R. Vranken, R. van der Mei, R. E. Kooij, and H. van den Berg. Performance of TCP with Multiple Priority Classes. Technical Report 279TD(02)010, COST-279, 2002.

[TD(02)011] M. Mandjes, D. Mitra, and W. Scheinhardt. Models of Network Access Using Feedback Fluid Queues. Technical Report 279TD(02)011, COST-279, 2002.

[TD(02)012] R. J. Boucherie and A. Ule. Adaptive Dynamic Channel Borrowing in Road-Covering Mobile Networks. Technical Report 279TD(02)012, COST-279, 2002.

[TD(02)013] M. Fiedler and G. Haßlinger. Waiting Time Quantiles for the Gaussian Voice Traffic Model. Technical Report 279TD(02)013, COST-279, 2002.

[TD(02)014] A. Medina, N. Taft, K. Salamatian, S. Bhattacharyya, and C. Diot. Traffic Matrix Estimation, State of Art and New Directions. Technical Report 279TD(02)014, COST-279, 2002.

[TD(02)015] K. Salamatian, B. Baynat, and T. Bugnazet. Cross Traffic Estimation by Loss Process Analysis. Technical Report 279TD (02)015, COST-279, 2002.

[TD(02)016] R. Litjens and H. van den Berg. Fair adaptive scheduling in integrated services UMTS networks. Technical Report 279TD (02)016, COST-279, 2002.

[TD(02)017] G. Latouche and M.-A. Remiche. A MAP-Based Poisson Cluster Model for Web Traffic. Technical Report 279TD(02)017, COST-279, 2002.

[TD(02)018] P. Salvador, A. Pacheco, and R. Valadas. Multiscale Fitting Procedure for Markov Modulated Poisson Processes. Technical Report 279TD(02)018, COST-279, 2002.

[TD(02)019] F. Delcoigne, A. Proutière, and G. Régnié. Modelling the Integration of Streaming and Elastic Traffic. Technical Report 279TD(02)019, COST-279, 2002.

[TD(02)020] I. Norros and H. Reittu. On Large Random Graphs with Infinite Variance Pareto Degree Distribution. Technical Report 279TD(02)020, COST-279, 2002.

[TD(02)021] M. Fiedler, P. Carlsson, K. Tutschku, S. Chevul, and A. A. Nilsson. Obtaining Reliable Bit Rate Measurement in SNMP-Managed Networks. Technical Report 279TD(02)021, COST-279, 2002.

[TD(02)022] J. Virtamo, E. Kuumola, and J. Resing. Joint Distribution of Instantaneous and Averaged Queue Length in an M/M/1/K System. Technical Report 279TD(02)022, COST-279, 2002.

[TD(02)023] J. van Leeuwaarden, S. Aalto, and J. Virtamo. Load Balancing in Cellular Networks Using First Policy Iteration. Technical Report 279TD(02)023, COST-279, 2002.

[TD(02)024] E. Hyytiä and E. Sirén. Delay Line Configurations in Optical Burst Switching with JET Protocol. Technical Report 279TD (02)024, COST-279, 2002.

[TD(02)025] P. Mannersalo. Some Notes on Prediction of Teletraffic. Technical Report 279TD(02)025, COST-279, 2002.

[TD(02)026] N. Akar, M. Atik, I. Dogru, and I. Hökelek. ATM Multipath Traffic Engineering Using Differentiated ABR. Technical Report 279TD(02)026, COST-279, 2002.

[TD(02)027] M. Ajmone Marsan, M. Franceschinis, E. Leonardi, F. Neri, and A. Tarello. Instability Phenomena in Underloaded Packet Networks with QoS Schedulers. Technical Report 279TD(02)027, COST-279, 2002.

[TD(02)028] M. Meo, M. Ajmone Marsan, and C. Batetta. Resource Management Policies in GPRS Wireless Internet Access Systems. Technical Report 279TD(02)028, COST-279, 2002.

[TD(02)029] W. Burakowski and H. Tarasiuk. Admission Control for TCP Connections in QoS IP Network. Technical Report 279TD (02)029, COST-279, 2002.

[TD(02)030] V. Inghelbrecht, B. Steyaert, S. Wittevrongel, and H. Bruneel. Analysis of the Interdeparture Process in Consecutive Stages of a VoIP Network. Technical Report 279TD(02)030, COST-279, 2002.

[TD(02)031] M. Menth and O. Rose. Performance Tradeoffs for Header Compression in MPLS Networks. Technical Report 279TD (02)031, COST-279, 2002.

[TD(02)032] E. Plasser. On the Non Linearity of the RED Drop Function. Technical Report 279TD(02)032, COST-279, 2002.

[TD(02)033] O. Østerbø. An Approximative Method to Calculate the Distribution of End-to-End Delay in Packet Networks. Technical Report 279TD(02)033, COST-279, 2002.

[TD(02)034] H. van den Berg, R. Kooij, M. Mandjes, P. Lassila, and R. van der Mei. Refinement of PS Models for TCP Performance. Technical Report 279TD(02)034, COST-279, 2002.

[TD(02)035] P. Gao, S. Wittevrongel, and H. Bruneel. Queueing Analysis of a Discrete-Time Multiserver Buffer System with Geometric Service Times. Technical Report 279TD(02)035, COST-279, 2002.

[TD(02)036] J. Walraevens, S. Wittevrongel, and H. Bruneel. Performance Evaluation of Optical Packet Switches Using Probability Generating Functions: an Initial Analysis. Technical Report 279TD (02)036, COST-279, 2002.

[TD(02)037] H. T. Tran and T. Ziegler. Engineering Solution of a CAC Mechanism for Voice Traffic over IP Networks. Technical Report 279TD(02)037, COST-279, 2002.

[TD(02)038] Ö. Özkasap and M. Çaglar. Traffic Properties of Scalable Multicast Communication. Technical Report 279TD(02)038, COST-279, 2002.

[TD(02)039] W. Burakowski and M. Fudala. PFS Scheme for Forcing Better Service in Best Effort IP Network. Technical Report 279TD (02)039, COST-279, 2002.

[TD(02)040] J. Kilpi. A Portrait of a GPRS/GSM Session. Technical Report 279TD(02)040, COST-279, 2002.

[TD(02)041] R. Susitaival, J. Virtamo, and S. Aalto. Load Balancing by MPLS in Differentiated Services Networks. Technical Report 279TD(02)041, COST-279, 2002.

[TD(02)042] K. Heck, D. Staehle, and K. Leibnitz. Diversity Effects on the Soft Handover Gain in UMTS Networks. Technical Report 279TD(02)042, COST-279, 2002.

[TD(02)043] F.-J. Ramón, J. Enríquez, J. Andrés, and A. Molínes. Multipath Routing with Dynamic Variance. Technical Report 279TD (02)043, COST-279, 2002.

[TD(02)044] M. Mohorčič, G. Kandus, and A. Švigelj. Traffic Flow Model for Routing Study in Packet-Switched Intersatellite Link Networks. Technical Report 279TD(02)044, COST-279, 2002.

[TD(02)045] R. Litjens and R. J. Boucherie. Performance Analysis of Downlink Shared Channels in a UMTS Network. Technical Report 279TD(02)045, COST-279, 2002.

[TD(02)046] K. Tutschku and M. Fiedler. Application of the Stochastic Fluid Flow Model for Bottleneck Identification and Classification. Technical Report 279TD(02)046, COST-279, 2002.

[TD(03)001] R. Litjens, H. van den Berg, R. J. Boucherie, F. Roijers, and M. Fleuren. Performance analysis of wireless LANs: An Integrated Packet/Flow Level Approach. Technical Report 279TD (03)001, COST-279, 2003.

[TD(03)002] K. Spaey, T. Hofkens, and C. Blondia. Timescales in Models for Bursty Traffic. Technical Report 279TD(03)002, COST-279, 2003.

[TD(03)003] R. Susitaival and S. Aalto. Providing Differentiated Services by Load Balancing and Scheduling in MPLS Networks. Technical Report 279TD(03)003, COST-279, 2003.

[TD(03)004] C. Blondia, O. Casals, L. Cerdà, N. van den Wijngaert, G. Willems, and P. de Cleyn. Low Latency Handoff Mechanisms and Their Implementation in an IEEE 802.11 Network. Technical Report 279TD(03)004, COST-279, 2003.

[TD(03)005] H. de Meer and K. Tutschku. A Measurement Study on Signaling in Gnutella Overlay Networks. Technical Report 279TD (03)005, COST-279, 2003.

[TD(03)006] J. Soler Lucas, V. B. Iversen, and M. Nord. Simulation and Performance Analysis of a GMPLS Lambda Scheduler. Technical Report 279TD(03)006, COST-279, 2003.

[TD(03)007] T. Bonald, P. Olivier, and J. Roberts. Dimensioning High Speed IP Access Networks. Technical Report 279TD(03)007, COST-279, 2003.

[TD(03)008] P. Salvador, A. Nogueira, and R. Valadas. Joint Characterization of the Packet Arrival and Packet Size Processes of Multifractal Traffic based on Stochastic L-Systems. Technical Report 279TD(03)008, COST-279, 2003.

[TD(03)009] S. Imre and M. Szalay. Reliability Modeling of Tree Topology Micro Mobility Networks. Technical Report 279TD(03)009, COST-279, 2003.

[TD(03)010] E. Karasan, N. Akar, O. Karasan, and M. Pinar. Mesh Topology Design in Overlay Virtual Private Networks. Technical Report 279TD(03)010, COST-279, 2003.

[TD(03)011] N. Akar and K. Sohraby. Numerically Stable Solution of Large-Scale Finite Fluid Queues. Technical Report 279TD(03)011, COST-279, 2003.

[TD(03)012] A. Baiocchi and A. de Vendictis. Investigating TCP Single Source Behavior in Time-Varying Capacity Network Scenarios. Technical Report 279TD(03)012, COST-279, 2003.

[TD(03)013] B. Zovko-Cihlar, W. Afric, and S. Grgic. Interference in Direct Spread Sequence Mobile Communication System. Technical Report 279TD(03)013, COST-279, 2003.

[TD(03)014] D. Z. Lenardic, B. Zovko-Cihlar, and M. Grgic. Analysis of Network Buffering Effects on TCP/IP Protocol Behavior. Technical Report 279TD(03)014, COST-279, 2003.

[TD(03)015] K. Laevens and H. Bruneel. Analysis of a Single-Wavelength Optical Buffer. Technical Report 279TD(03)015, COST-279, 2003.

[TD(03)016] I. Norros and H. Reittu. On the Architecture of the Power-Law Random Graph Model of Internet. Technical Report 279TD (03)016, COST-279, 2003.

[TD(03)017] D. Fiems, S. de Vuyst, and H. Bruneel. Discrete-Time Analysis of the Gated-Exhaustive Vacation Queue. Technical Report 279TD(03)017, COST-279, 2003.

[TD(03)018] D. Staehle and A. Mäder. An Analytic Approximation of the Uplink Capacity in a UMTS Network with Heterogeneous Traffic. Technical Report 279TD(03)018, COST-279, 2003.

[TD(03)019] S. Köhler and A. Binzenhöfer. MPLS Traffic Engineering in OSPF Networks—A Combined Approach. Technical Report 279TD(03)019, COST-279, 2003.

[TD(03)020] J. Karvo and S. Aalto. Using Multicast or a Combination of Unicast and Broadcast for Transmitting Popular Content. Technical Report 279TD(03)020, COST-279, 2003.

[TD(03)021] M. Fiedler, K. Tutschku, P. Carlsson, and A. A. Nilsson. Identification of Performance Degradation in IP Networks Using Throughput Statistics. Technical Report 279TD(03)021, COST-279, 2003.

[TD(03)022] R. Litjens, H. van den Berg, and R. J. Boucherie. Throughput Measures for Processor Sharing Models. Technical Report 279TD(03)022, COST-279, 2003.

[TD(03)023] J. Kilpi. Distributional Properties of GPRS/GSM Session Volumes and Durations. Technical Report 279TD(03)023, COST-279, 2003.

[TD(03)024] A. Švigelj, M. Mohorčič, G. Kandus, A. Kos, M. Pustišek, and J. Bešter. Consideration of Empirical IP Traffic in Traffic Class Dependent Packet-Switched ISL Routing. Technical Report 279TD(03)024, COST-279, 2003.

[TD(03)025] S. Imre. Dynamically Optimised Chernoff Bound Based CAC for 3G/4G WCDMA Systems. Technical Report 279TD(03)025, COST-279, 2003.

[TD(03)026] N. Benameur, S. Oueslati, and J. W. Roberts. Experimental Implementation of Implicit Admission Control. Technical Report 279TD(03)026, COST-279, 2003.

[TD(03)027] N. Akar, I. Hökelek, M. Atik, and E. Karasan. A Reordering-Free Multipath Traffic Engineering Architecture for DiffServ-MPLS Networks. Technical Report 279TD(03)027, COST-279, 2003.

[TD(03)028] I. Gojmerac, T. Ziegler, F. Ricciato, and P. Reichl. Adaptive Multipath Routing for Efficient Load Balancing in the Internet. Technical Report 279TD(03)028, COST-279, 2003.

[TD(03)029] M. Menth, J. Charzinski, and S. Kopf. Impact of Resilience Requirements on the Performance of Network Admission Control Methods. Technical Report 279TD(03)029, COST-279, 2003.

[TD(03)030] G. Hu, K. Dolzer, and C. Gauger. Does Burst Assembly Really Reduce the Self-Similarity. Technical Report 279TD(03)030, COST-279, 2003.

[TD(03)031] S. Szabó, S. Imre, and A. Burulitisz. On the Accuracy of Mobility Modelling in Wireless Networks. Technical Report 279TD (03)031, COST-279, 2003.

[TD(03)032] L. Muscariello, M. Mellia, M. Meo, R. Lo Cigno, and M. Ajmone Marsan. A Simple Markovian Approach to Model Internet Traffic at Edge Routers. Technical Report 279TD(03)032, COST-279, 2003.

[TD(03)033] D. Sass, K. Dolzer, S. Bodamer, and M. Lorang. Analytic Fluid Flow Approach for Fair Queueing Systems. Technical Report 279TD(03)033, COST-279, 2003.

[TD(03)034] H. van den Berg, M. Mandjes, R. van de Meent, A. Pras, F. Roijers, and P. Venemans. QoS Aware Bandwidth Provisioning in IP Backbone Networks. Technical Report 279TD(03)034, COST-279, 2003.

[TD(03)035] W. Burakowski and A. Beben. Premium Message Handling Service in IP QoS Networks. Technical Report 279TD(03)035, COST-279, 2003.

[TD(03)036] Ö. Özkasap and M. Çaglar. Traffic Characterization of Scalable Multicasting in the Case of a Self-Similar Source. Technical Report 279TD(03)036, COST-279, 2003.

[TD(03)037] V. Inghelbrecht, B. Steyaert, and H. Bruneel. Burstification Mechanism of an OBS Edge Router: Analytical Analysis of a One- and Two-Threshold Case. Technical Report 279TD(03)037, COST-279, 2003.

[TD(03)038] U. Krieger, V. Klimenok, L. Breuer, and A. N. Dudin. The BMAP/G/1 Queue with Feedback Operating in Synchronous Random Environment as Model for a Telecommunication Channel with Performance Fluctuation. Technical Report 279TD (03)038, COST-279, 2003.

[TD(03)039] U. Krieger and N. Markovitch. On-Line Estimation of Heavy-Tailed Traffic Characteristics in Web Data Mining. Technical Report 279TD(03)039, COST-279, 2003.

[TD(03)040] J. Charzinski, C. Hoogendoorn, K. Schrodi, C. Winkler, and M. N. Huber. Towards Carrier-Grade Next Generation Networks. Technical Report 279TD(03)040, COST-279, 2003.

[TD(03)041] P. Gao, S. Wittevrongel, and H. Bruneel. Queueing Analysis of Multiserver Buffers with Geometric Service Times and Correlated Input Traffic. Technical Report 279TD(03)041, COST-279, 2003.

[TD(03)042] P. Carlsson, M. Fiedler, and A. Ekberg. On an Implemention of a Distributed Passive Measurement Infrastructure. Technical Report 279TD(03)042, COST-279, 2003.

[TD(03)043] K. Tworus, S. de Vuyst, S. Wittevrongel, and H. Bruneel. Queueing Analysis of the Stop-and-Wait ARQ Protocol in a Wireless Environment. Technical Report 279TD(03)043, COST-279, 2003.

[TD(03)044] W. Burakowski and M. Dabrowski. Assessment of Token Bucket Parameters by On-Line Traffic Measurements. Technical Report 279TD(03)044, COST-279, 2003.

[TD(03)045] A. Beben and R. Janowski. Statistical Admission Control for WWW Traffic. Technical Report 279TD(03)045, COST-279, 2003.

[TD(03)046] M. Menth, J. Milbrandt, and A. Reifert. Backup Capacity Minimization for Simple Protection Switching Mechanisms. Technical Report 279TD(03)046, COST-279, 2003.

[TD(03)047] H. van den Berg, A. I. Endrayanto, and R. J. Boucherie. Characterizing CDMA Downlink Feasibility via Effective Interference. Technical Report 279TD(03)047, COST-279, 2003.

[TD(03)048] M. Mandjes, P. Mannersalo, I. Norros, and M. van Uitert. Most Probable Busy Period Paths in Gaussian Queues. Technical Report 279TD(03)048, COST-279, 2003.

[TD(03)049] K. Tutschku and P. Tran-Gia. A Traffic Profile of the eDonkey Filesharing Service. Technical Report 279TD(03)049, COST-279, 2003.

[TD(03)050] Ö. Özkasap. Scalability and Robustness of Pull-Based Anti-Entropy Distribution Model. Technical Report 279TD(03)050, COST-279, 2003.

[TD(04)001] S. Imre and M. Szalay. LTRACK — Novel Location Management Method. Technical Report 279TD(04)001, COST-279, 2004.

[TD(04)002] L. Isaksson, M. Fiedler, and A. Nilsson. Validation of Simulations of Bluetooth's Frequency Hopping Spread Spectrum Technique. Technical Report 279TD(04)002, COST-279, 2004.

[TD(04)003] A. Kortebi, S. Oueslati, and J. Roberts. Cross-Protect: Implicit Service Differentiation and Admission Control. Technical Report 279TD(04)003, COST-279, 2004.

[TD(04)004] A. Švigelj, M. Mohorcic, and G. Kandus. Analysis of Different Scheduling Policies in Traffic Class Dependent Routing for LEO Satellite Networks with ISLs. Technical Report 279TD(04)004, COST-279, 2004.

[TD(04)005] I. Norros and H. Reittu. On a Conditionally Poissonian Graph Process. Technical Report 279TD(04)005, COST-279, 2004.

[TD(04)006] K. Salamatian, A. Soule, and N. Taft. Flow Classification by Histograms or How to Go to Safari Over Internet. Technical Report 279TD(04)006, COST-279, 2004.

[TD(04)007] A. de Vendictis, F. Vacirca, and A. Baiocchi. TCP over Wireless Links with Rayleigh Fading: Some Considerations about Channel Energy Efficiency and TCP Performance. Technical Report 279TD(04)007, COST-279, 2004.

[TD(04)008] R. Khalili and K. Salamatian. On the Distribution of Error Events Length in Convolutional Codes. Technical Report 279TD (04)008, COST-279, 2004.

[TD(04)009] K. Salamatian and S. Fdida. A Framework for Interpreting Measurement Over Internet. Technical Report 279TD(04)009, COST-279, 2004.

[TD(04)010] A. Binzenhöfer and P. Tran-Gia. Delay Analysis of a Chord-based Peer-to-Peer File-Sharing System. Technical Report 279TD (04)010, COST-279, 2004.

[TD(04)011] R. Martin, M. Menth, and V. Phan-Gia. Performance of TCP/IP with MEDF Scheduling. Technical Report 279TD(04)011, COST-279, 2004.

[TD(04)012] S. Molnár, T. A. Trinh, and B. Sonkoly. A Performance Study of High Speed TCP. Technical Report 279TD(04)012, COST-279, 2004.

[TD(04)013] D. Rossi, L. Muscariello, and M. Mellia. Analyzing TCP Flow Arrivals. Technical Report 279TD(04)013, COST-279, 2004.

[TD(04)014] T. Bonald and A. Proutière. Insensitive Bandwidth Sharing in Data Networks. Technical Report 279TD(04)014, COST-279, 2004.

[TD(04)015] M. Mandjes, P. Mannersalo, I. Norros, and M. van Uitert. Large Deviations of Infinite Intersections of Events in Gaussian Processes. Technical Report 279TD(04)015, COST-279, 2004.

[TD(04)016] M. Dabrowski, A. Beben, and W. Burakowski. On Inter-Domain Admission Control Supported by Measurements in Multi-Domain IP QoS Network. Technical Report 279TD(04)016, COST-279, 2004.

[TD(04)017] M. L. Garcia Osma, F.-J. Ramón Salguero, G. Garcia de Blas, J. Andrés Colás, and J. Enríquez Gabeiras. Enabling Local Preference in Peer-to-Peer Traffic. Technical Report 279TD(04)017, COST-279, 2004.

[TD(04)018] T. Maertens, J. Walraevens, and H. Bruneel. Analysis of Priority Queues with Priority Jumps. Technical Report 279TD(04)018, COST-279, 2004.

[TD(04)019] P. Gao, S. Wittevrongel, K. Laevens, and H. Bruneel. Dicrete-Time Multiserver Preemptive Resume Priority Queues with Geometric Service Times. Technical Report 279TD(04)019, COST-279, 2004.

[TD(04)020] O. Alparslan, N. Akar, and E. Karasan. Combined Use of Prioritized AIMD and Flow-Based Traffic Splitting for Robust TCP Load Balancing. Technical Report 279TD(04)020, COST-279, 2004.

[TD(04)021] H. van den Berg, R. Litjens, and J. Laverman. HSDPA Flow Level Performance: the Impact of Key System and Environment Aspects. Technical Report 279TD(04)021, COST-279, 2004.

[TD(04)022] M. Bottigliengo, C. Casetti, C.-F. Chiasserini, and M. Meo. Short-term Fairness for TCP Flows in 802.1lb WLANs. Technical Report 279TD(04)022, COST-279, 2004.

[TD(04)023] L. Breuer, G. Latouche, and M.-A. Remiche. An EM Algorithm for the Model Fitting of a Class of Transient Markovian Arrival Processes. Technical Report 279TD(04)023, COST-279, 2004.

[TD(04)024] F. Roijers, X. Fan, and H. van den Berg. Analysis of TCP Flow Transfer Times Over IEEE 802.11 Wireless LANs. Technical Report 279TD(04)024, COST-279, 2004.

[TD(04)025] R. Pries and K. Heck. Performance Comparison of Handover Mechanisms in Wireless LAN Networks. Technical Report 279TD (04)025, COST-279, 2004.

[TD(04)026] T. Hoßfeld and D. Staehle. Semi-Analytic Model of the UMTS Downlink Capacity with WWW Traffic on Dedicated Channels. Technical Report 279TD(04)026, COST-279, 2004.

[TD(04)027] L. Isaksson and M. Fiedler. Optimization of the Random Backoff Boundary of the Bluetooth Frequency Hopping Spread Spectrum Technique. Technical Report 279TD(04)027, COST-279, 2004.

[TD(04)028] W. Burakowski, J. Sliwinski, and A. Beben. An Approach for Effective Handling of Constant Bit Rate Traffic Over WLANs. Technical Report 279TD(04)028, COST-279, 2004.

[TD(04)029] E. Hyytiä, P. Lassila, L. Nieminen, and J. Virtamo. Spatial Node Distribution in the Random Waypoint Mobility Model. Technical Report 279TD(04)029, COST-279, 2004.

[TD(04)030] R. Janowski and K. Rzepakowski. Estimation of the Admissible Load in a Two Priority System. Technical Report 279TD(04)030, COST-279, 2004.

[TD(04)031] N. Akar and E. Karasan. Exact Calculation of Blocking Probabilities for Bufferless Optical Burst Switched Links with Partial Wavelength Conversion. Technical Report 279TD(04)031, COST-279, 2004.

[TD(04)032] E. Hyytiä and J. Virtamo. Random Waypoint Model in n-Dimensional Space. Technical Report 279TD(04)032, COST-279, 2004.

[TD(04)033] A. de Vendictis, F. Vacirca, and A. Baiocchi. Experimental Analysis of TCP and UDP Traffic Performance over Infra-Structured 802.11b WLANs. Technical Report 279TD(04)033, COST-279, 2004.

[TD(04)034] D. Fiems, K. Laevens, and H. Bruneel. Analysis of a Two-Stage Optical Buffer. Technical Report 279TD(04)034, COST-279, 2004.

[TD(04)035] J. L. Sobrinho, R. de Haan, and J. M. Brázio. Why RTS-CTS Is Not Your Ideal Wireless LAN Multiple Access Protocol. Technical Report 279TD(04)035, COST-279, 2004.

[TD(04)036] S. De Vuyst, S. Wittevrongel, and H. Bruneel. A Queueing Discipline with Place Reservation. Technical Report 279TD(04)036, COST-279, 2004.

[TD(04)037] V. B. Iversen and T. Holmberg. Resource Sharing Models for Quality-of-Service. Technical Report 279TD(04)037, COST-279, 2004.

[TD(04)038] M. Franceschinis, M. Mellia, M. Meo, and M. Munafo. Measuring TCP over WiFi: A Real Case. Technical Report 279TD (04)038, COST-279, 2004.

[TD(04)039] Z. Genç and Ö. Özkasap. Peer-to-Peer Epidemic Algorithms for Reliable Multicasting in Ad Hoc Networks. Technical Report 279TD(04)039, COST-279, 2004.

[TD(05)001] M. Klimo and K. Bachratá. Impact of a Packet Loss Process to a Speech Process. Technical Report 279TD(05)001, COST-279, 2005.

[TD(05)002] R. Litjens, H. van den Berg, and M. J. Fleuren. Spatial Traffic Heterogeneity in HSDPA Networks and its Impact on Network Planning. Technical Report 279TD(05)002, COST-279, 2005.

[TD(05)003] F. Vacirca, T. Ziegler, and E. Hasenleithner. Large Scale Estimation of TCP Spurious Timeout Events in Operational GPRS Networks. Technical Report 279TD(05)003, COST-279, 2005.

[TD(05)004] M.-A. Remiche. Compliance of the Token Bucket Model with Markovian Traffic. Technical Report 279TD(05)004, COST-279, 2005.

[TD(05)005] P. Arlos and M. Fiedler. A Comparison of Measurement Accuracy for DAG, Tcpdump and Windump. Technical Report 279TD(05)005, COST-279, 2005.

[TD(05)006] A. Kortebi, L. Muscariello, S. Oueslati, and J. Roberts. Evaluating the Number of Active Flows in a Scheduler Realizing Fair Statistical Bandwidth Sharing. Technical Report 279TD(05)006, COST-279, 2005.

[TD(05)007] S. Aalto and U. Ayesta. Mean Delay Comparison among Multilevel Processor Sharing Scheduling Disciplines. Technical Report 279TD(05)007, COST-279, 2005.

[TD(05)008] T. Hoßfeld, K. Tutschku, and F.-U. Andersen. Mapping of File-Sharing onto Mobile Environments: Enhancement by UMTS. Technical Report 279TD(05)008, COST-279, 2005.

[TD(05)009] T. Maertens, J. Walraevens, and H. Bruneel. Analysis of a GI-1-1 Queue with a Modified HOL Priority Scheduling Discipline. Technical Report 279TD(05)009, COST-279, 2005.

[TD(05)010] M.-I. Jeannin, L. Bernaille, A. Soule, and K. Salamatian. Blind Applicative Flow Recognition Through Behavioral classification. Technical Report 279TD(05)010, COST-279, 2005.

[TD(05)011] R. Khalili and K. Salamatian. A Routingless Approach for Communication in Wireless Networks. Technical Report 279TD (05)011, COST-279, 2005.

[TD(05)012] W. Rogiest, K. Laevens, D. Fiems, and H. Bruneel. Analysis of an Asynchronous Optical Buffer. Technical Report 279TD(05)012, COST-279, 2005.

[TD(05)013] J. Sliwinski, A. Beben, and W. Burakowski. Approach for Effective Handling of CBR Traffic over WLANs: Heterogeneous CBR Sources. Technical Report 279TD(05)013, COST-279, 2005.

[TD(05)014] H. Tarasiuk, W. Burakowski, and R. Janowski. On Assuring End-to-End QoS in Heterogeneous Networks by Investigating Network Service Concept. Technical Report 279TD(05)014, COST-279, 2005.

[TD(05)015] A. Alagoz, E. Ahi, and Ö. Özkasap. Network Awareness and Buffer Management in Epidemic Information Dissemination. Technical Report 279TD(05)015, COST-279, 2005.

[TD(05)016] S. Molnár and T. A. Trinh. Analysis of TCP Vegas and FAST TCP in a Game-Theoretical Framework. Technical Report 279TD (05)016, COST-279, 2005.

[TD(05)017] I. Norros and H. Reittu. On the Performance and Stability of Peer-to-Peer Networks with Highly Variable Life Times of Nodes. Technical Report 279TD(05)017, COST-279, 2005.

[TD(05)018] B. Zovko-Cihlar and I. Milak. Digital Video Processing in Wireless Multimedia Transmissions. Technical Report 279TD (05)018, COST-279, 2005.

[TD(05)019] N. Sengezer and E. Karasan. TSCP: A Tabu Search Algorithm for Wavelength Converting Node Placement in WDM Optical Networks. Technical Report 279TD(05)019, COST-279, 2005.

[TD(05)020] R. Martin, M. Menth, and J. Charzinski. Comparison of Border-to-Border Budget Based Network Admission Control and Capacity Overprovisioning. Technical Report 279TD(05)020, COST-279, 2005.

[TD(05)021] M. Menth, J. Milbrandt, and S. Kopf. Adaptive Bandwidth Allocation for Wide Area Networks. Technical Report 279TD(05)021, COST-279, 2005.

COST 279 Management Committee

Austria

Dr. Thomas Ziegler
Forschungszentrum Telekommunikation Wien
ziegler@ftw.at

Belgium

Prof. Herwig Bruneel
Universiteit Gent
herwig.bruneel@telin.ugent.be

Prof. Guy Latouche
Université Libre de Bruxelles
latouche@ulb.ac.be

Croatia

Prof. Branka Zovko-Cihlar
University of Zagreb
branka.zovko@fer.hr

Cyprus

Prof. Soulla Louca
Intercollege—Dept. of Computer Science
louca@intercollege.edu

Dr. Andreas Pitsillides
University of Cyprus
Andreas.Pitsillides@ucy.ac.cy

Denmark

Prof. Villy Baek Iversen
Danish Technical University—COM Center
vbi@com.dtu.dk

Finland

Prof. Ilkka Norros
VTT Information Technology
ilkka.norros@vtt.fi

Prof. Jorma Virtamo
Helsinki University of Technology—Networking Laboratory
jorma.virtamo@hut.fi

France

Mr. Philippe Olivier
France Telecom
phil.olivier@francetelecom.com

Prof. Kavé Salamatian
University of Paris VI
salamat@rp.lip6.fr

Germany

Prof. Udo Krieger
Otto-Friedrich Universität Bamberg
udo.krieger@ieee.org

Prof. Phuoc Tran-Gia
Universität Würzburg
trangia@informatik.uni-wuerzburg.de

Hungary

Dr. Sándor Imre
Budapest University of Technology and Economics
imre@hit.bme.hu

Dr. Sándor Mólnar
Budapest University of Technology and Economics
molnar@ttt-atm.bme.hu

Italy

Prof. Andrea Baiocchi
Università di Roma "La Sapienzia"
baiocchi@infocom.uniroma1.it

Dr. Michela Meo
Politecnico di Torino
michela@polito.it

Netherlands

Dr. Hans van den Berg
TNO Telecom
j.l.vandenberg@telecom.tno.nl

Dr. Remco Litjens
TNO Telecom
r.litjens@telecom.tno.nl

Norway

Dr. Terje Jensen
Telenor Research & Development
terje.jensen1@telenor.com

Dr. Mette Rohne
Telenor Research & Development
mette.rohne@telenor.com

Poland

Prof. Wojciech Burakowski
Warsaw University of Technology
wojtek@tele.pw.edu.pl

Portugal

Prof. José Brázio
Telecommunications Institute/IST
jose.brazio@lx.it.pt

Prof. Rui Valadas
Telecommunications Institute/UAveiro
rv@av.it.pt

Serbia and Montenegro

Ms. Mirjana Stojanovic
Mihailo Pupin Institute
stojmir@kondor.imp.bg.ac.yu

Slovakia

Prof. Martin Klimo
University of Zilina
Martin.Klimo@fri.utc.sk

Slovakia (cont.)

Prof. Pavol Podhradsky
Faculty of Electrical Engineering and Information Technology
podhrad@ktl.elf.stuba.sk

Slovenia

Prof. Gorazd Kandus
Josef Stefan Institute
gorazd.kandus@ijs.si

Spain

Prof. Olga Casals[†]
Polytechnic University of Catalonia

Mr. Manuel Villén Altamirano
Telefónica I+D
manolo@tid.es

Sweden

Dr. Markus Fiedler
Blekinge Institute of Technology
markus.fiedler@bth.se

Dr. Adrian Popescu
Blekinge Institute of Technology
adrian@nada.kth.se

Turkey

Prof. Nail Akar
Bilkent University
akar@ee.bilkent.edu.tr

UK

Dr. Yim-Fun Hu
Leeds Metropolitan University
f.hu@lmu.ac.uk

Dr. Peter Key
Microsoft Research Ltd
peterkey@microsoft.com

[†] Deceased June 2003.

COST 279 Participating Institutions

Austria
Telecommunications Research Center Vienna

Belgium
Ghent University
Université Libre de Bruxelles
University of Antwerp

Canada
Royal Military College of Canada

Croatia
Croatian Telecom
KPMG Consulting
University of Zagreb

Cyprus
Intercollege—Deparment of Computer Science
University of Cyprus

Denmark
Technical University of Denmark

Finland
Helsinki University of Technology
Tampere University of Technology
VTT Information Technology

France

École Nationale Supérieure des Télécomunications—Paris
France Telecom
University Paris VI

Germany

Deutsche Telekom T-Nova
Institute of Communication Networks, TU München
Otto Friedrich University Bamberg
Siemens, AG
University of Stuttgart
University of Würzburg

Hungary

Budapest University of Technology and Economics

Italy

Politecnico di Torino
Università di Roma "La Sapienza"

Netherlands

Center for Mathematics and Computer Science
Delft University
TNO Telecom
University of Twente

New Zealand

University of Canterbury

Norway

Telenor Research & Development

Poland

Warsaw University of Technology

Portugal

Instituto Superior Técnico
Telecommunications Institute
University of Aveiro

Serbia and Montenegro

Mihailo Pupin Institute

Slovakia

Slovak University of Technology

University of Zilina

Slovenia

Jozef Stefan Institute

Spain

Universitat Politecnica de Catalunya

Telefónica I&D

Sweden

Blekinge Institute of Technology

Ericsson Core Network Development

Lund University

Telia Research AB

Turkey

Bilkent University

Koç University

United Kingdom

Leeds Metropolitan University

Microsoft Research Ltd